Machine Sensation

New Metaphysics

Series Editors: Graham Harman and Bruno Latour

The world is due for a resurgence of original speculative metaphysics. The New Metaphysics series aims to provide a safe house for such thinking amidst the demoralizing caution and prudence of professional academic philosophy. We do not aim to bridge the analytic-continental divide, since we are equally impatient with nail-filing analytic critique and the continental reverence for dusty textual monuments. We favor instead the spirit of the intellectual gambler, and wish to discover and promote authors who meet this description. Like an emergent recording company, what we seek are traces of a new metaphysical 'sound' from any nation of the world. The editors are open to translations of neglected metaphysical classics, and will consider secondary works of especial force and daring. But our main interest is to stimulate the birth of disturbing masterpieces of twenty-first century philosophy.

Tessa Leach

Machine Sensation: Anthropomorphism and 'Natural' Interaction with Nonhumans

O
OPEN HUMANITIES PRESS
London, 2020

First edition published by Open Humanities Press 2020
Copyright © 2020 Tessa Leach
Freely available online at http://openhumanitiespress.org/books/titles/machine-sensation/

This is an open access book, licensed under a Creative Commons By Attribution Share Alike license. Under this license, authors allow anyone to download, reuse, reprint, modify, distribute, and/or copy this book so long as the authors and source are cited and resulting derivative works are licensed under the same or similar license. No permission is required from the authors or the publisher. Statutory fair use and other rights are in no way affected by the above. Read more about the license at creativecommons.org/licenses/by-sa/3.0

Design by Katherine Gillieson
Cover Illustration by Tammy Lu

The cover illustration is copyright Tammy Lu 2020, used under a Creative Commons By Attribution license (CC-BY).

PRINT ISBN 978-1-78542-082-5
PDF ISBN 978-1-78542-081-8

OPEN HUMANITIES PRESS

Open Humanities Press is an international, scholar-led open access publishing collective whose mission is to make leading works of contemporary critical thought freely available worldwide. More at http://openhumanitiespress.org

Contents

Introduction   9
    Anthropomorphism   12
    The value of taking nonhuman perspectives seriously   19
    A note on language   28
    Guiding questions   33
    Structure of this book   40

1. ANT and OOO for technology   45
    Actor-Network Theory (ANT)   46
    Object-Oriented Ontology (OOO)   49
    Alien phenomenology   56

2. Kinect and "Natural" Interfaces   67
    The objects themselves   70
    The Kinect-brand   79
    Kinect-artefact 1.0   84
    Kinect-artefact 2.0   87
    Communication, sensation and relation of the Kinect-artefact   91
    Kinect Sports Rivals   100
    Alien phenomenology of the Kinect-artefact   107

3. Grey Walter's Tortoises and an Introduction to the Sensation and Experience of Robots   115
    Umwelten and cybernetics   117
    Robot tortoises   124
    Sensation, experience and consciousness   131

4. SHRDLU and Bidirectional Postphenomenology   144
    What is (or was) SHRDLU?   146
    Postphenomenology   157
    SHRDLU and Ihde's categories   169
    Consequences of and problems with reversing
    Ihde's relation types in SHRDLU   180
    Micro- and macroperception   183

5. Gynoids and the Politics of the Alien   186
    Artificial women are inherently political   190
    Notable qualities of the concept of gynoid sexbots   197
    Human investigations into the concept of gynoid sexbots   211
    Gynoid sexbot power in ANT and OOO   215

Conclusion   225
    A broadening of the concepts of sensation and experience   226
    Technological relations beyond anthropomorphism   231

Notes   237

Bibliography   267

List of Figures

Figure 1: Kinect-artefact 2.0's relationship to other objects

Figure 2: Necker Cube

Rather than wondering if alien beings exist in the cosmos, let's assume that they are all around us, everywhere, at all scales: not just dogs and penguins and magnolia trees, but also cornbread and polyester and the city of Orlando and sidewalks and nectarines. I started asking one particular question about their inhumanity, a question I have previously given the name *alien phenomenology*: what do objects experience? What is it like to be a thing?

        Ian Bogost, "Inhuman," in *Inhuman Nature*

Introduction

The Voyager 2 spacecraft, currently hurtling at great speed away from Earth and towards the edge of our familiar solar system, along with its twin Voyager 1, will be the first human-made machines to pass beyond the heliopause, and to encounter the objects that make up interstellar space.[1] Sometime after this, perhaps as early as 2025, its sensors will be switched off and it will continue, inert, through space until it approaches the star Sirius in about 296,000 years.[2] The Voyager spacecraft are not the first human *objects* to cross into interstellar space – human-made electromagnetic waves began leaving Earth over a century ago. But the Voyagers are the most anthropomorphic object yet to leave behind the aura of our sun's energy, the energy that is the root of life on Earth, of human life, and ultimately of human technology. Voyager 2 is a machine built by humans, incorporating metals extracted by human ingenuity from the Earth, plastic derived from decayed organic matter and plutonium manipulated by highly skilled human engineers. There is also a time capsule aboard the vessel, carrying information about the human race considered relevant by the scientists who launched it in 1977.[3] Among other things, this includes a disc known as the "Golden Record", containing analogue-coded photographs of human life, greetings in 55 different Earth languages, and music from Beethoven to Chuck Berry. Moreover, it carries sensory organs designed to gather information about phenomena that human scientists consider important, and it conveys that information in a way that humans on Earth can receive and interpret. Voyager 2 is an object made of human materials and ideas, and it is careering as fast as possible away from us, making its way further into alien territory by the instant.

The point of this story is not to marvel at the way in which an artificial object, intrinsically linked to science and human exploration, can gradually become alien to us. In popular parlance, the alien is that which is strange, inhuman and sometimes hostile. It is often used to describe the possible residents of other planetary bodies, perhaps beyond the edge of our solar system. Our desire to explore will soon bring a human-made artefact in contact with this alien territory. But Voyager 2 was *already* an alien before it was ever launched into space, despite the immense human effort put into creating it. Like all nonhumans, the qualities that define Voyager 2 and its way of being in the world are for the most part inaccessible to us and are evident only in interactions between it and other objects – that is, how it senses and relates to its environment (including us), and how its environment relates to it. We built this machine, yet the argument that it is not fully accessible to us is a key aspect of one of the most important theoretical influences on this book, Graham Harman's object-oriented philosophy (OOP), also known as object-oriented ontology (OOO). OOO is a prominent set of theoretical, practical and artistic ideas that have significantly influenced the so-called nonhuman turn, a movement in the contemporary humanities away from an anthropocentric view of the world and towards foregrounding nonhumans and their perspectives. In later pages, the details of Harman's ontology and other similar positions will be described, depicting the universe as a glittering fractal of objects and their relations. An object like Voyager 2 has complex relationships both with individual humans (such as engineers and politicians) and with a human culture that promotes exploration of the universe, but it also interacts with billions of nonhuman objects (such as electrons and stars) in ways that are closed off from our experience.

Nevertheless, something about Voyager 2 will reflect its human origins. Its means of signalling to Earth, its evidence of human workmanship and particularly its cargo of precious images and ideas, all point to its close relations with humanity that may well persist beyond the lifespan of the human race. Voyager 2 possesses human-like qualities in that it bears distinct evidence of close modelling upon humans and human intent in their design, although it is unlikely to be confused with a human. In contrast, the case studies discussed in this book are frequently compared with humans. We project humanness onto them and pick and choose the alien qualities that are of most interest to us. But this book will argue that every anthropomorphic machine, every sensor, every robot, every interface, every word, and every concept of future or transformative technology is as alien to us as is Voyager 2.

Anthropomorphic machines are marvellous. We marvel at them. From intuitive chatbot interfaces to fully-fledged bipedal robots, we love to look for human-like qualities in machines. Anthropomorphic machines are charming when they work and frustrating when they fail. This is not necessarily because we see ourselves as the epitome of good engineering; we do not need to think of ourselves as quite that vain. But in many machines, particularly those that we bring into our homes, anthropomorphism aids in our ability to use them since humans are typically very good at communicating with other humans. They usually reproduce elements of human bodies or performance that make it easier for human users to comprehend how they relate to the world, apparently replicating qualities like eyesight and the ability to learn to speak. But we do not build machines to be like humans. We build machines to *pretend* to be humans. And by discussing specific technologies alongside contemporary anthrodecentric philosophy, this book aims to address the marvellous complexity of machine sensation and experience that anthropomorphic discourse tends to obscure. To a large extend, this is done through the mobilisation of and experimentation with OOO.

In engineering contexts, "anthropomorphic machines" typically refers to bipedal or social robots. Here, the term is much broader. I leave it deliberately vague in acknowledgement of the fact that anthropomorphism can be seen in an enormous variety of technological objects and other nonhumans. But a key factor that links all of the case studies in this book is the fact that they were designed or marketed as material or metaphorical mimics of humans. The case studies are, in the order presented:

1. The Microsoft Kinect, a natural user interface (NUI) for the popular Microsoft Xbox One (a videogame console);

2. William Grey Walter's robot tortoises (*Machina speculatrix*);

3. Terry Winograd's SHRDLU, an early language-using computer program; and

4. The concept of the gynoid sex robot.

These case studies were selected for particular reasons, and those reasons will be explained in each respective chapter. Although these objects are all human-made, and all possess qualities that make them anthropomorphic, I argue that anthrodecentric inquiry focusses investigation onto how anthropomorphic qualities in machines may obfuscate and confuse questions about their place in human society. Each are examples of protagonists in episodes when humans and artefacts have managed to comprehend each other successfully enough to form a useful working relationship, as well as occasions when an artefact stubbornly refuses to cooperate, when the human expects more than the artefact can deliver, or when the human does not act in a way that the artefact requires in order to do what the human wants it to do. Anthropomorphism is something that is done by humans in reaction to particular qualities in machines. Not everyone sees the same anthropomorphic qualities in machines. What might seem human-like to one human may to another human seem uncanny or a tool through which governments or corporations pursue their own agendas. This is an obstacle to consensus-building about the possible societal consequences of building anthropomorphic machines. This book contends that these episodes are usefully illuminated through greater exploration of the artefacts' sensory worlds and speculation about their internal experiences, even if those experiences are radically separate from us. In other words, we should take nonhuman perspectives seriously.

Anthropomorphism

Since it is easy for us to interact with human-like objects, and it is often easier to copy living things rather than invent brand new ways of being in the world, humans frequently anthropomorphise advanced machines.

Anthropomorphic qualities can improve the performance of artefacts, or they can improve human knowledge of human bodies and minds. The machines discussed in this book are all anthropomorphised for at least one of these two reasons. But anthropomorphism is a trick that is deceptive but persuasive, and consequently valuable or dangerous depending on the context.[4]

We might think of technology as existing on a spectrum of complexity. One end of this complexity spectrum is occupied by the simplest of machines, such as levers and glass cubes, and the other is occupied by our most advanced computers. Inevitably, the advanced end of the spectrum has become associated with machines that we heavily anthropomorphise. Computers "think", "hear" and "make mistakes". These complex machines are the subject of much study both in philosophy and cultural studies. They have directed us to rethink what it means to be human and how we interact with objects. They visibly, and often deliberately, blur boundaries that modern Westerners are used to thinking of as impervious.

But our tendency to anthropomorphise is also evident in other kinds of objects. It can be seen in the assignment of human qualities to objects in the natural world, such as gods, animals and natural disasters.[5] Moreover, we can trace human interest in anthropomorphic nonhumans through stories of more primitive machines. The humanoid robots found in science fiction are the modern incarnations of venerable stories like Pygmalion's statue and the golem (a clay humanoid enchanted into movement by a rabbi).[6] Narratives in which human-like sculptures come to life have been common since ancient times and are represented in early science fiction, when robots were monstrous and eerie.[7] Medieval automata and Japanese *karakuri* were frequently made in the image of humans,[8] and today the compulsion is evident in our most sophisticated AI systems. The case studies of this book are diverse in structure and sensory capabilities, but they share similarities because of the human-like attributes that we give them.

The desire for anthropomorphism often comes into conflict with requirements that we may have for machines, such as skills that humans do not have. It also sometimes diverts attention to things that do not aid in the task assigned to the nonhuman. As Brian Duffy puts it, "anthropomorphism obstructs the fact that human form is not the ideal for a machine."[9] But that does not necessarily mean that designing human-like characteristics into a machine is necessarily frivolous or vain. Anthropomorphism in technology is a practical concern (such as when designing robots for therapeutic purposes[10]) as well as being indicative of certain preoccupations of human designers (theorists have identified preoccupations such as vanity as well as religious faith[11]). Anthropomorphised entities are more likely to be deemed worthy of our care and consideration.[12] For example, the term "Mother Earth" is often used by people who wish others to behave in an environmentally conscious way.[13] Studies have shown that humans are more likely to deem nonhumans responsible for their actions if the nonhumans have anthropomorphic qualities.[14] The greater the potential action and "intelligence" of the technology, the greater we consider its ability for rational thought and decision-making. Claiming that an artefact is human-like can also be a way of emphasising its dangerous qualities. One of the most common tropes in science fiction is that of robots who are created to help humans but end up posing a threat. They are almost always anthropomorphic robots. They might be visually anthropomorphic, like the robots gone amok in *Westworld*,[15] or they might have functionally anthropomorphic language and conversation skills like HAL in *2001: A Space Odyssey*.[16] It is psychologically easier for humans to furnish objects with moral responsibility if they have a stronger resemblance to humans.

As Kate Darling points out, anthropomorphic framing of technological artefacts may be desirable or undesirable (from a human point of view).[17] Anthropomorphising an object allows us to manipulate the response of human consumers, which might be a good thing with a Fitbit that encourages people to exercise more, but a negative thing if it can persuade people to release personal data.[18] The resemblance need not be physical. One experiment by Clifford Nass and colleagues revealed that humans are even anxious about offending computers.[19] Participants performed a task on Computer A and were then asked to evaluate the computer's performance on the same machine. But a second group was asked to enter their opinion of Computer A on Computer B. When the participants evaluated the computer on the same machine, participants were less critical than when the evaluation was performed on Computer B. As Sherry Turkle puts it, "participants do not want to insult a computer 'to its face.'"[20] Even when machines are not designed to look human, people impose human values and relationships onto them. Anthropomorphism can be useful for designers who wish to emphasise the human-like qualities of their creations for profit.

Machines may only bear a resemblance to part of the human form, and that part is likely emphasised by the designer. Take the case of a glass eye. In this case the goal is to replicate as realistically as possible the appearance of the eye, without its visual functions. But another type of anthropomorphism is evident in technologies that mimic the function of human bodies. A pertinent corollary to the glass eye is the retinal prosthesis, a technology used in the treatment of macular degeneration, a common eye condition that can seriously impede human vision. The retinal prosthesis attempts to merge naturally with the remaining eye tissue and stimulate parts of the retina.[21] It simulates human sensation, but it is structurally very dissimilar from the eye and the user is visibly cyberised. The glass eye is visually anthropomorphic, while the retinal prosthesis is functionally anthropomorphic. Functional anthropomorphism can also be seen in our daily interaction with technology: in the assignment of names to our computers, or the impulse to swear at them when they malfunction.[22]

In semi-autonomous technological artefacts, sensation and perception of external data may be more or less functionally anthropomorphic. For example, a robot arm in an assembly line needs to be able to sense any instructions being sent to it from an external object (such as a computer), to have proprioception (an understanding of the position of its "body parts" in space) and to have a knowledge of where other objects are located (such as the object being assembled). Other machines have sensory apparatuses that make them better suited to environments with a lot of people. A social robot is most effective with both visual anthropomorphism and functional sensory anthropomorphism. It should be able to relate what it sees, hears, feels and smells in a way that seems natural to human users. Yet when we consider the mechanisms that produce these effects, it is evident that natural appearances are in fact highly contrived artificial senses. Machines do not have ears, nor do they have the centres of the brain responsible for hearing, but may have microphones and sound chips instead. Their sensation and perception of phenomena begins as machine-like and is heavily mediated to become human-like. We can have some insight into how the sound chip functions, easily performing some feats and being unable to perform others. But ultimately the structure of the sound card is only related to us through other objects. Why would we choose functionally anthropomorphic or visually anthropomorphic qualities in different cases? Our choice reveals the intended use and cultural or practical value of the artefact.

Choosing when to use anthropomorphic language and when to use more alien language can divide us on social issues. The boundary between human and nonhuman is drawn, policed, and transgressed in each of the case studies discussed. Acceptance of anthropomorphism comes more easily to some people than others. To some humans, anthropomorphic machines are endearing. To others, the same qualities that are meant to be endearing seem uncanny or even blasphemous.[23] Sex robots (or sexbots) are discussed in Chapter 5 of this book and are an excellent example. The ethics of acting out a rape fantasy with a sexbot is dependent on the perceived anthropomorphic or alien qualities of the sexbot. To an individual who sees more alien qualities in a sexbot, it is more tool-like than human-like and it cannot be raped. To an individual who sees more anthropomorphic qualities in a sexbot, the question of consent becomes more relevant. Note that this may not be because the sexbot is deserving of rights for her own sake, but because violence against machines with anthropomorphic qualities could lead to violence against human beings. Gender and ethnic identity are also difficult to define in robots because of the conflict between perceived human and alien qualities. Darling cites the power of anthropomorphism to both reinforce and challenge stereotypes.[24] Nonhumans that lack anthropomorphic qualities like ATMs and factory robots cannot be accused of representing particular gendered or ethnic groups. But if a nonhuman is highly anthropomorphic then gendered and ethnic signifiers *must* be built into them. As discussed in Chapter 2, even a simple machine can be accused of racism or sexism. Seeing more of a human or more of an alien in a nonhuman determines the degree to which it is interpolated into cultural and social issues. Anthropomorphism is subjective, but examining an object's alien qualities alongside their human ones can help us to make more informed decisions about what they "understand" or "desire". We could make an informed decision about the ethics of sexbot rape if we could access the inner being of individual sexbots, but we cannot. All we can do is focus on the relations they form with both physically embodied and incorporeal objects. It can be tempting to focus on anthropomorphic qualities in nonhumans that approach the carefully policed human/nonhuman divide. But to be anthrodecentric we must focus on the whole object and accept our psychological impression of it as just one side of the story.

It might not be possible to know exactly what a pencil is experiencing. But it is not possible to know what another human is experiencing either, yet we still give voice to each other and advocate on behalf of marginalised groups. Anthropomorphic machines are built in our image – are they a marginalised group? Historically we have designated groups as less than human in order to justify violence or subjugation. Rosi Braidotti has called "the human… a normative convention, which does not make it inherently negative, just highly regulatory and hence instrumental to processes of exclusion and discrimination."[25] Braidotti is speaking here not of machines, but of groups of humans that have been historically excluded from the concept of humanity (such as enslaved peoples). But we still regulate humanity and either resist or encourage the exclusion of anthropomorphic machines from human categorisation. How human-like does an object have to be for us to call it human? If we insist upon anthropomorphising large groups of nonhumans, we are met with a contradiction. We frequently refuse to consider them as individuals. When anthropomorphic machines are shipped en masse, we simultaneously inflict humanity on them while refusing them the individuality that humans, according to Western morality, deserve.

Of course, there is a problem with this argument in that it equates the suffering and struggle of marginalised or oppressed groups of humans with the human labelling of machines. Being dehumanised is a big problem for a human or groups of humans. But if anthropomorphic machines are marginalised by us, it is not necessarily important to them. Who can say what a social robot really *wants* – perhaps it wants to be a good companion to humans; perhaps it wants to encourage us to build more social robots; perhaps it wants to maintain access to its electric power source; or perhaps it simply wants to exist for as long as possible. We cannot be sure. If a social robot tells me it is happy, I am inclined to disbelieve it. But there is a certain point on the human-nonhuman spectrum at which we need to take seriously what nonhumans tell us about their experiences. Because it has anthropomorphic qualities, and was in part created by us in our own image, it is unconscionable to avoid acknowledging its social robot-ness. If we do not at least attempt to create an anthrodecentric narrative for the social robot, then how can we justify bringing these machines into the world? We deny them self-determination by avoiding their alien qualities. We contrive human-like qualities then force them into an unhuman outgroup. Nonhumans do not necessarily mind being forced into an outgroup, but *humans* should mind. We have enough trouble appreciating existing humans' individuality without creating more stereotyped anthropomorphic outgroups.

The challenge of examining relations in anthropomorphic machines is often complicated by cyborg or hybrid qualities in objects. It is becoming harder to determine where human agency, decision-making and sensation ends and where that of nonhumans begins. When we can't tell what is human and what is nonhuman, the distinctions between them become less relevant in an expedient sense, even if they are still philosophically interesting. Thus we are left with interesting hybrids, actor-networks, and cyborgs: the employment of algorithms to buy and sell in that bastion of human culture and economics, the Stock Exchange; the use of artificial steroids to enhance athletes that participate in the Olympics, once the test of "natural" human bodies; the close touch of a robot, controlled by a surgeon outside the room, in the removal of cancerous cells.[26] The case studies in this book are deliberately presented as human-like, not just as human-machine hybrids. The chapters ahead contain examples of genuine attempts by humans to create an artificial other, often with personalities, names and the attribution of human qualities. Once again, these objects can best be interrogated by examining their relations. An essential first step is to decide upon the object of study and not be distracted by our own sloppy language choices. Is it the robot or robot-surgeon network that is the focus of investigation? Are we talking about the Stock Exchange as an embodied network, or as an incorporeal concept?

Another reason why anthropomorphic machines make interesting case studies is the fact that they are generally explicable by their creators.

Because anthropomorphic machines are engineered by humans, we can make useful deductions about their way of being in the world. In Chapter 3 the robot tortoises discussed possessed simple light and touch sensors. These sensors were built into them. Unlike when we study nonhumans such as bats and slime moulds, the question of which parts of the world the nonhuman is capable of sensing is already partly answered. In most cases, creators are able to understand why technological artefacts act in certain ways.[27] There's no need to reverse engineer a sensory organ if you've created it. In addition to this, the sensory organs of anthropomorphic machines can often be compared to human sensory organs. A light sensor may attempt to replicate human vision, but we know enough about the anatomy of the eye and the design of the sensor to deduce that they have different structures and, consequently, a different way of being in the world. Comparing the two yields an appreciation of what it is that makes nonhumans unlike humans and more alien. We can generalise this appreciation to apply to all nonhumans.

The value of taking nonhuman perspectives seriously

This book encounters anthropocentric thought on two fronts. The anthropomorphism of machines is the first. The second is the anthropocentric thought that dismisses the sensation and experience of nonhumans beyond that of their successful or unsuccessful interaction with human "users". Both of these tendencies distort nonhumans into specific kinds of caricatures. We are inhibited in our attempts to look beyond human problems or thinking, which can be a hindrance when persuading technological objects to behave one way or another. They may spring surprises on us. So, this book proposes to employ an anthrodecentric approach both to the theory and the case studies, a technique that actively removes the human from the centre of scholarly inquiry. We may discover which anthropomorphic qualities we give to (or inflict on) machines. An uninhibited approach to studying nonhuman phenomenological worlds can open our minds and, with any luck, facilitate more informed relationships between technological artefacts and humans. There is irony in this: the deployment of anthrodecentric thought to solve anthropocentric problems.

Science and philosophy are both heavily invested in the work of describing the lived experiences of humans. Yet there has been relatively little work studying the phenomenal world of the machine as a semi-autonomous being. The language to explain, for example, what constitutes the experience of a spacecraft, microphone, length of copper wire or videogame character, is still in its infancy (possibly due to a lack of interest in any aspect of its experience beyond that which is relevant to its technological functionality). This book argues that this is a significant problem for the philosophy of technology because it is only by comprehending the phenomenology of machines that we can arrive at an understanding of experiences between humans and nonhumans (or between nonhumans).

To begin this book, I must make one thing clear: no human can ever fully understand what it is like to be a robot or a computer. According to Harman, objects are radically divided from one another and the only part of the objects we can see are those that are evident in our interaction with them. I once believed that OOO could help increase sympathy between humans and anthropomorphic machines, which might smooth a path toward a more congenial relationship. But over time I have come to think that the opposite goal is also a worthy task. Anthropomorphism already promotes a feeling of sympathy for a nonhuman, but that sympathy sometimes obscures the alien processes that contribute to unanticipated failures in human-artefact relations. We sometimes underestimate the importance of paying attention to other relations such as those with materials, light, platforms, and ideas. OOO permits (and even encourages) a loss of sympathy through extreme alienation. But it can also spark interest in nonhuman experiences, and particularly in the sensory capabilities that nonhumans use to interact with one another.

The version of OOO presented in this book is based upon Harman's work, from *Tool-Being: Heidegger and the Metaphysics of Objects* (which is based on his 1999 doctoral dissertation) to more recent publications. Throughout his career, Harman's metaphysical position has remained steady, but it has found new areas of application and he and other authors have allowed OOO to flourish in unexpected ways. There have been other interpretations of a OOO nature, but I focus on Harman because of his prominent status in the movement and his continuing observation of and engagement with the changes taking place. The evolution of his writing is significant. His first books were deeply concerned with metaphysics and particularly phenomenology. But more recent works have addressed the socio-political implications of OOO.

Harman's metaphysics is rich and vast. It is built on an interpretation of Heidegger that leads him to some unconventional conclusions.[28] Harman's concepts of the quadruple object, withdrawal, and allure form a robust structure through which the interpretation of real examples is made possible. They provide the language to facilitate accounts of complex relations between objects. For the purposes of this book, it is not important whether Harman is correct in his ontology because it is mainly used here to focus the anthrodecentric inquiry into the case studies. Harman's model provides a cohesive framework for studying specific objects and is a useful tool for both purely metaphysical and socio-political inquiry.

Chapter 1 will go into detail about OOO, but to those who are unfamiliar with the idea here is a quick contextualisation. OOO is fundamentally different from most other approaches to theorising being and relations because of its anti-correlationism. The word "correlationism" in this context was first used by Quentin Meillassoux in 2006 in his book *After Finitude*.[29] It refers to the prevailing understanding in philosophy that the human view of the world is the only one that is possible or important to understand. All other perspectives are either inaccessible to humans or are not worth considering. Levi Bryant identifies a profound anthropocentrism at the heart of correlationism.[30] However, most post-Kantian philosophers, from phenomenologists to idealists, are correlationists. In contrast, OOO is an anti-correlationist movement based on realism. It insists not only that nonhumans really exist outside of human perception but that we can know something of objects independent of human access. As Ian Bogost says, "humans are allowed to live […] alongside sea urchins, kudzu, enchiladas, quasars, and Tesla coils."[31] Each of these nonhumans is an object, but the definition also extends to humans and to incorporeal things like concepts and signs. And each object, no matter its humanlike or corporeal status, is equally ontologically relevant. The perspective of the sea urchin is just as necessary for an understanding of the world as that of the human, even if it does not seem important to most humans when compared with our own perspectives. Matt Hayler points to OOO's "usefulness" in the study of technological artefacts in particular.[32] Our bodies limit and structure our encounters with the world, and technologies both change our encounters with the world and exist as independent and inaccessible objects in their own right. Anti-correlationist thought is often a key concept for those interested in the concept of "flat ontology",[33] since anti-correlationism discourages hierarchies and categories of objects. At the time of writing, Bogost and cultural critic Christopher Schaberg were commissioning books and essays on "the hidden lives of ordinary things", such as golf balls and refrigerators.[34] In general, these texts on ordinary things follow the trend of rejecting the primacy of human lives and experience in favour of a focus on nonhumans, a project that is anti-correlationist in nature.

OOO is a metaphysics of objects in general, and narrowing the focus to anthropomorphic machines is already a step away from pure metaphysics toward the socio-political implications of theory. OOO has important consequences for the study of technology beginning with its problematisation of both STEM disciplines (Science, Technology, Engineering and Maths) and HASS disciplines (Humanities, Arts and Social Sciences). Each of these paradigmatic approaches to the investigation of nonhumans are traditionally and inherently correlationist. Harman analyses each of these approaches in a short essay entitled *The Third Table*, and he has different objections to each of them that amount to the same criticism: both the STEM and HASS disciplines enable the "reality" of things to disappear from inquiry and each "reduces" objects in different ways.[35] (This is related to the concepts of overmining, undermining, and duomining, which are discussed in Chapter 1.) The first approach critiqued in *The Third Table* is the investigation of objects through the natural sciences. When Voyager 2 is investigated through the STEM disciplines, we look at its material qualities. Of what materials is it built, and how do those materials cause it to react to different natural forces like gravity and electricity? How do those materials and their interactions affect human usage? Different parts of Voyager 2 are there to perform different functions, such as the sensors, transmitters, and the provision of power to the machine. This study of materials, if taken far enough, leads us to consider the atomic and sub-atomic structure of different parts of Voyager 2. Scientific investigation deflects attention from the object as a whole. Voyager 2 is not just a set of atoms "any more than a game is just a set of plays or a nation just a set of individuals."[36] On the other hand, societal and cultural investigation into objects such as those made by the social sciences also avoid studying the object for its own sake and from its own perspective. In the case of Voyager 2, the social sciences might study its economic impact, the historical factors leading up to specific engineering choices made about its form, why as a species we feel the need to incorporate a record of our culture and society in a machine destined for interstellar drifting, and whether those records encoded on the Golden Record were Euro-centric in nature. These investigations, in Harman's view, also reduce the object albeit to "its theoretical, practical, or causal effects on humans or on anything else."[37] The object itself is avoided. The object itself is "an intermediate being found neither in subatomic physics nor in human psychology, but in a permanent autonomous zone where objects are simply themselves."[38] Voyager 2 may have been created by the manipulation of material structures and ideas, but there also exists a Voyager 2 that is irreducible to any of these things. There is a Voyager 2 that exists for its own sake. There is a lack of interest in this Voyager 2-ness that OOO can target. Even if we can't access the nonhuman, acknowledging the alien status of Voyager 2 gives us a fresh starting point.

So much of human experience is affected by technology, so our interest in it tends to be self-centred and limited. Engineering and cultural texts about anthropomorphic technologies report on an inner world in a way that reduces the tension between the alien experiences of the nonhuman and its positioning in human culture. Typical questions include whether it is possible to build a machine that is genuinely intelligent or ethical. This question would involve reference to both STEM and HASS disciplines. To be clear, Harman is not advocating the abandonment of STEM or HASS investigations. These are ways that humans gain knowledge about the world, and are important.[39] But these kinds of investigations often marginalise the alien qualities of nonhumans by bringing them under the aegis of human-centred research. We reduce the tension between humans and nonhumans in a way that is not always in our best interest. Ironically, understanding of the perspective of nonhumans allows us to advance human interests. One way of doing this is using OOO and alien phenomenology.

One of the most important tenets of OOO is the rejection of human-centric enquiry or the common *a priori* belief that objects are only important because of their relationship with humans, a concept that it shares with actor-network theory (ANT). Voyager 2, drifting beyond our ability to retrieve it on its long voyage to Sirius, is swiftly proceeding towards a space uninhabited by humans – popularly known as the territory of E.T. and little green men. This is not the intended usage of the term "alien" here. "Alien" in this book is much closer to Ian Bogost's description of alien phenomenology (see Chapter 1).[40] The word is used to highlight the failure of human language or understanding to describe the unhuman processes, experiences and tendencies of nonhumans. It need not refer to human-made or complicated nonhumans. Tectonic plates are alien. Aerosol sprays are alien. Smartphones are, as Lev Manovich says in his comments on the creation and customisation of human-computer interfaces, "friendly aliens."[41] We can never feel what it is like to be one of these nonhumans. But that should not be our goal (except perhaps in artistic practices). Rather we should investigate the questions that come up when we explore the capacity of anthropomorphic technologies to exist in worlds that may not centre around humans. Once Voyager 2 finally moves out of range of Earth, it will still exist as much as it did before. Humans are not the centre of the universe, and human observation does not define nonhumans. The star Sirius (and any nearby inhabitants) is unlikely to register any kind of human presence for nearly 300,000 years.

OOO and alien phenomenology are linked more broadly to the nonhuman turn in the humanities, and it is a metaphysical orientation that is of interest to any movement that aims to decentre the human. The nonhuman turn may be traced to such ideas as ANT, affect theory, animal studies, assemblage theory, brain sciences and artificial intelligence, new media theory, new materialism and systems theory.[42] The nonhuman turn also incorporates different theoretical approaches. It is an extremely varied movement, but the term "nonhuman turn" articulates a certain direction in ideas which can be traced back to the Copernican revolution. Upon discovering that the Earth revolved around the sun, the anthropocentric universe was suddenly decentred. The Earth was no longer the centre of the universe, surrounded by circling planets, a moon and a sun. The Copernican revolution resulted in a conclusion that the sun was at the centre of the universe, and that we revolve around it. As recounted by Bruce Mazlish, this was a major blow for the human conception of self.[43] Mazlish goes on to say that the second blow came with Darwin, and the work that so significantly altered science and culture. No longer were humans a separate entity, perhaps created by a god, fundamentally ruptured from all other animals. Humans had evolved from animals. Evolutionary theory was around before Darwin, but he was among the first to apply the concept to human evolution in *The Descent of Man*. A third major blow came with Freud and the psychoanalytic revolution. Freud argued that the human mind is not entirely the domain of the human. Everybody has an ego, the mode in which we live most of the time, but with a considerable controlling influence by our superego and id. The id was characterised as being somewhat animalistic and primal, implying that the human mind is not actually as rational as we might think.

More recently there has been an increasing awareness of the dire ecological danger posed by, among other things, climate change. Most discourse on this topic is directed towards the protection of humans from this terrible calamity. But some ideas on this topic have resulted in a decentring of the human and a more posthuman perspective. The work of James Lovelock is a particularly good example of this. In his book *Gaia: A New Look at Life on Earth*, Lovelock reframes the planet as a sort of ecological agent with its own homeostatic system, in which life is a co-creator of the Earth's conditions.[44] He anthropomorphises the planet as Gaia, after the Ancient Greek Earth goddess.[45] Gaia is capable of responding to the presence of human beings the way a human would respond to a virus. The humans change the climate of the planet, which kills us off, solving the problem from Gaia's perspective.[46] He also draws the interesting comparison between human pollution and the creation of oxygen by photosynthetic bacteria, aeons ago, destroying the anaerobic microorganisms in an "oxygen pollution disaster".[47] This implies that while humans might be worried about the conditions for human life on Earth, the Earth itself is neutral about ecological destruction: "The very concept of pollution is anthropocentric and it may even be irrelevant in the Gaian context."[48] This is an example of an ecological perspective resulting in an orientation toward the nonhuman. The Anthropocene is the name that is increasingly being given to the period of geological time that we are currently living in, a word premised on the fact that human beings are making such a profound difference to the ecological make-up of the planet that it may be recognised as a distinctive time period.[49] It is not officially recognised by geological experts, although some advocate the use of the term, some estimating that it began with the industrial revolution and others dating it back further to the start of agriculture. Qualities of the Anthropocene include mass extinction, extreme alteration of the air, land and aquatic systems, and the deposit of heavy metals. The concept of the Anthropocene is occasionally cited as a motivating factor for participants in the nonhuman turn, since ecological narratives are so frequently motivated by an over-riding concern for the welfare of human beings over, for instance, coral reefs, bees and forests. Timothy Morton has called the Anthropocene an "antianthropocentric concept" that forces us to face "the task of thinking at temporal and spatial scales that are unfamiliar, even monstrously gigantic."[50] We are faced with catastrophes that are simultaneously of our own making and that are yet ludicrously removed from us in terms of scale.

Some theorists have become disenchanted with a philosophical model of the universe that prioritises the human. Rejecting the Kantian model in which observation by a human is central to ontology, Harman, Ray Brassier, Iain Hamilton Grant and Quentin Meillassoux briefly formalised a new anthrodecentric direction in philosophy under the umbrella of "Speculative Realism".[51] I say briefly because the name was adopted in April 2007 before speculative realism splintered into radically different movements.[52] Harman, Brassier, Hamilton Grant and Meillassoux now agree on little.[53] OOO is the principal movement that will be mobilised in this book, with frequent references to Harman. There is little critique of OOO metaphysics here. (Many already exist – see for example Slavoj Žižek's comments,[54] which primarily critique Bryant's ontology; Alexander Galloway's comments,[55] which criticise what he sees as Harman's apolitical attitude towards ontology; and Nathan Brown's comments,[56] which condemn OOO as obscurantism). Instead it is hoped that OOO and other anthrodecentric philosophies can bring about greater awareness of machine phenomenology.

Machine phenomenology is not a trivial matter, as we will discover when the first data on interstellar plasma is interpreted by human scientists. At that time, the relationship between the plasma and the human-made spacecraft will be of great interest, and the rendering of its observations into human terms will depend on our understanding of Voyager 2's ability to mediate new knowledge for the human eyes, ears and brains on Earth. Typically, the output of data and changes in the state of the object are the only observations available to us in interrogating an artefact, and only a skilled observer can distinguish the useful observations from the irrelevant ones. A robot might relate to me by directing its eyes towards my face and speaking, and this is the focus of my experience, but we must not forget that it is also in a relationship with the floor, photons, code and the materials of which it is made, and they are all contributing to its experience at that time.

As humans we cannot access the inner Voyager 2-ness of Voyager 2. We can only explore the relations that it forms with other objects (including humans), and our view of it is most complete when we take a transdisciplinary approach. Alien phenomenology should not just be used to investigate relations between nonhumans, but also between specific objects of study and human authors. Each author takes different things from their investigation into objects – including authors with backgrounds in the natural sciences and the social sciences (and including the present author). The trick is in taking these authors and their conclusions as objects in themselves, which also form part of the flat ontology. No perspective on objects should be taken as a definitive perspective, since the only real perspective is the one possessed by the object, which is inaccessible to us. All investigations are partial investigations, and no matter their disciplinary background, they are equally objects in relation to the alien under investigation. Alien phenomenology should be transdisciplinary. Social and scientific studies may co-exist peacefully in an alien phenomenological approach, since the origins of the study are not important to the alien. All relations between humans and nonhumans are influenced by the ideological, disciplinary and cultural history of the human, as well as by physical qualities such as the possession of sensory organs. With a OOO analysis, these qualities of human authors are impossible to avoid, but each is equally valid. This doesn't mean that exploration of nonhumans is pointless or impossible. Each contribution to knowledge around a nonhuman adds to a collective alien phenomenology of the nonhuman. Starting with acknowledgement of the alien in objects like anthropomorphic machines decentres human inquiry, and permits a broader critical study of these objects. This is both an advantage and a limitation of OOO. We can arrive at a more complete investigation into nonhumans, but only if we incorporate multiple approaches.

This book makes a conscious effort to increase the tension between nonhuman and human experience by emphasising the alien qualities of anthropomorphic technologies which are so often neglected by humans. The discourse surrounding technology is anthropocentric, as is the way in which humans speak about specific technological artefacts. The discourse centres around human-nonhuman relations and cultural impact, and we speak of (and to) technological artefacts as if they were humans. Hence, this book confronts the same issue on two fronts. We need to *accentuate* the anthropomorphic in the machine in order to promote their alien faculties. Humans frequently conceal or overlook aspects of machine phenomenology because of the mundane use of anthropomorphic language. HAL speaks to the crew, the automaton plays the piano, the robot opens her mouth. I will accentuate this tendency as much as possible so as to make this day-to-day anthropomorphic language stand out. Ian Bogost would call this a point "where gears grind"[57] and his example of the werewolf is relevant here. There is a friction between human and inhuman that is emphasised in the transformation, which we too often replace "with explanation or ignorance."[58] This book will not seek to erase the power of the transformation between anthropomorphic and nonhuman, between human and robot, between a robot that is switched on and one that is switched off. The transformation between these two states is what gives these entities their incredible power and is an indelible part of their objecthood.

A note on language

> As Samuel Butler says in the *Book of Machines*, 'Won't it be the glory of machines that they can do without the great gift of speech? Someone has said that silence is a virtue that makes us agreeable to our fellows.' Ah, Aramis, you would

> finally enjoy that silence. Why would you waste your time speaking? What you aspire to is not bearing the 'I.' On the contrary, your dignity, your virtue, your glory, lie in being a 'one.' And it is this silence, this happy anonymity, this depth, this heaviness, this humanity, that we have denied you. I am speaking in your place, I am offering you the awkward detour of prosopopoeia, but it is precisely because you are dead forever. 'It' wanted to become not the subject of our discourse, but the object, the tender anonymous object by means of which we would travel in Paris. Is that so hard to understand?
>
> Professor Norbert H. on the doomed Parisian transport system Aramis. Bruno Latour, "Epilogue," *Aramis or, the Love of Technology*, trans. Catherine Porter (Cambridge, MA: Harvard University Press, 1996), 297.

As Bruno Latour puts it, we are required to take the detour of prosopopoeia, the act of speaking as though on behalf of things that are abstract, absent, or dead. We do not have the language to avoid correlationist statements. The field of evolutionary biology may give us a clue to how to approach this issue. In biology, as in robotics, the question of teleology often arises. Living things are the result of millions of years of evolution by natural selection, with no external agency controlling our progress. For this reason, biologists use a form of shorthand in describing the structure and function of living things. This is known as teleonomy, or the practice of speaking of things as though they have a purpose while taking it for granted that in evolution by natural selection there is no purpose or objective.[59] For example, it is commonly said that the purpose of molar teeth is to grind down food to make it easier to digest. In his 1888 essay on the subject, Henry Fairfield Osborn describes the evolutionary transition of Mammalian molars from those of prehistoric phyla as a "reduction of primitive elements towards special adaptation."[60] In reality, molar teeth are the result of random mutation and selection with no direction or intent. In contrast, a cheese grater is designed by a human to break up food. It has a genuine teleology, not just the appearance of one. A similar practice is proposed in this book.

It is necessary to use somewhat anthropocentric language in order to make accessible statements about nonhumans. There are always limitations to a text: word limits, time limits, the author's knowledge, etc. An exhaustive evaluation of just one of Voyager 2's sensors would require several volumes, if we are to examine everything from its subatomic particles to its impact on human culture. This book employs a technique that we might call anthroponomy. It makes a statement about a nonhuman's sensation and experience using language that is usually used to describe human sensation and experience.

The author is human, the format is human-centric, and language itself is a human mode of relating to other objects. As soon as we use language to try to understand the experience of a nonhuman, we are anthropomorphising it. This book routinely states that certain technological artefacts "sense", "perceive", "experience" and "remember". Nonhumans have "intentions", "qualities" and "desires". Human experience is not special relative to nonhuman experience, so I have no hesitation in hijacking our specialised anthropocentric psychological and ethological language for an appreciation of nonhuman and non-living phenomenology. Chapter 3 goes further and addresses panpsychism, the idea that every object in the universe has some aspect of mind, consciousness or soul. But for the most part, this anthropocentric language is merely a product of our lack of useful shorthand for what occurs within non-living things. There is no shorthand language describing nonhuman perspective. But, as Laura Gustafsson and Terike Haapoja point out, this lack of non-anthropocentric language is not a good excuse for not making an attempt at writing about nonhuman experiences.[61] In their 2013 Helsinki exhibition entitled *History According to Cattle*, Gustafsson and Haapoja endeavour to approximate cattle-like experiences through language intended to evoke responses in humans and assist in conceptualising the alien worlds of nonhumans:

> Although we share many basic experiences with other species, abstract linguistic expression is, as far as we know, a uniquely human aptitude – yet language still remains hopelessly inadequate at conveying anything about corporeality or corporeal experience. Language nevertheless provides the human species with a mental toolkit for making sense of the

world; ideally it can serve as a bridge to the experiential realm of the Other.[62]

In a similar spirit, this book attempts a written exploration of technological experiences with only the linguistic tools available to cultural critics, philosophers, engineers and ethologists (all of whom are humans).

In many cases the creator or distributor of technologies has encouraged this use of metaphor. The Kinect, for example, is "smart". It "sees", "hears" and "thinks". Many technological artefacts have been given a gender-specific pronoun and are deemed to have particular qualities that align with human qualities, such as stubbornness, stupidity, attractiveness, kindness, love, and introspection. This book uses this language consistently while acknowledging that such human qualities are the consequences of alien processes inside the nonhuman. Although we clearly lack the linguistic tools to describe nonhuman experience, this should not stop us from trying. Wherever possible, the focus is on maintaining consistency throughout this book in the use of ethological language, theoretical terms and social conventions like pronouns.

There are a multitude of academic texts written in acknowledgement of the nonhuman turn, and many employ their own terminologies. I do not intend to debate which wordings are best, since I am interested in specific objects and not a purely metaphysical argument. The word "object" will be used, despite its baggage, to refer to all of the technological artefacts, ideas, living things and phenomena referenced. This book follows Harman and other object-oriented ontologists in calling all these things objects no matter their material status. The word "actant" may also be used, as in ANT, but this term does not emphasise the inherent passivity of objects that is suggested by Harman's metaphysics. The word "qualities" will be used in place of other synonyms, again because of the word's usage in Harman's writing. Harman describes both "sensual qualities" (qualities related to the sensual objects that allow interaction between objects) and "real qualities" (qualities that belong to the infinitely withdrawn object itself, but that are nevertheless divided from objects). (Chapter 1 will have more to say on this topic.) Unless otherwise specified, the word "qualities" in this book refers to real qualities, which are concealed from the view of any other object.

The use of the term "the human" is also a problematic one that I will nevertheless employ as shorthand. "The human" is not an object. Individual humans are objects, and human collectives are objects. But "the human" describes a set of qualities that are not necessarily universal in humans. The human has eyes, but not all humans have eyes. The human uses language, but not all humans use language. The human reproduces sexually, but not all humans reproduce sexually. It is important to note this caveat at the beginning of this work, not only to indemnify the author from claims of insensitivity but also to establish some object-oriented credentials early on. It is not tenable to support a universal view of humans. "Humanity" is quite another thing. "Humanity" is a concept, and it obeys the rules of concepts, changing over time and space in response to cultural influence. Humanity is a moralising term and sometimes a philosophical term, and consequently enters into relations with other objects. There is a difference between humanity and the human.[63]

Gendered language is occasionally used by me and others when describing humans and nonhumans. The gynoids described in Chapter 5, for example, are often given the pronoun "she" both by academics and by the public. The robots have been given qualities that enable them to perform a female identity, including certain linguistic markers. I have chosen to use this approach in recognition of the performative nature of gender, and of the attempts that are being made to erase the difference between these robots and humans (and humans are usually assigned a particular gender at birth). When it comes to humanoid robotics and AI, the aims of engineers and participants in the nonhuman turn are allied. Gendered pronouns are sometimes even used with simple robots such as Elsie and Elmer, simple robots described in Chapter 3. Created in the 1940s by W. Grey Walter, Elsie and Elmer possess little to mark them as female or male other than their names and the pronouns used to discuss them, but since this may also be true of humans I have used the same pronouns as Walter.

Finally, a comment on the use of personal pronouns with reference to myself and the text. Frequently, this book avoids the author's personal pronouns ("I", "me", "myself") depending on context. This stylistic choice is less common these days, since texts in the humanities depend upon the author's theoretical and cultural milieu and are subjective. However, the choice to refer to the text rather than the author is relevant to the subject matter of this book. I am not the same object as the book. The book contains information that I do not store in my body, and my body contains information not included in the book. A book is not an aspect of the author, or a medium through which they communicate; it is its own object with its own qualities, relations and agency. I will change over time in ways that the book cannot. It may be subjective, but it is not me. When I do use personal pronouns, I hope that the reader will forgive me and understand that I am using it for clarity or elegance of language.

Guiding questions

There are three main questions that help to regulate the application of theory to the technological objects of study in this book. The questions are not necessarily answered, but they are used to guide the discussion of the specific case studies.

How do we ask machines about their experiences?

This is a methodological question. What techniques can we use to learn about the phenomenology of nonhumans? The personal assistance agent Siri, who is a popular feature of Apple products, is known for giving witty responses to questions about her experience of the outside world. Obviously, understanding the phenomenology or sensory world of machines must go beyond such human-engineered approaches. We need a phenomenological survey of the object, performed in the same way that an ethologist would survey the phenomenological profile of a tick (see Chapter 3). In this book similar tactics are used; various questions must be asked and interactions pursued to encourage the object to show more of its qualities to the outside. Strongly associated with this question is "how does the phenomenology of an anthropomorphic technology affect its agency?" ANT is one way to approach this kind of research, and one that will be more familiar than OOO to most readers. ANT allows researchers to answer questions about phenomena of interest by examining the agency of all actants involved, both human and nonhuman. Latour, a key figure in ANT, sees an understanding and integration of nonhumans to be essential to the comprehension of any kind of scientific enquiry, since scientists must inevitably compel nonhumans to behave in a particular way in laboratory or field settings.[64] But Latour was originally a sociologist, and although his examples of laboratories and politics ground anthrodecentric ideas in empirical studies, ANT lacks the metaphysical strength of OOO. Latour himself jokes: "I'm like a dog following its prey, and then the prey arrives in the middle of a band of wolves which are called professional philosophers."[65]

The trouble with trying to ask machines about their experiences is that it must be done in a humanish way. It is impossible to fully study any other kind of relationship, since the author of this book is a human. But every effort must be made to remove the human from the centre of the analysis. Human concerns must be considered for these analyses, but they must not be privileged over other kinds of relations such as those between nonhuman and nonhuman. As Bryant puts it:

> The human-object relation is not a special relation, not a unique relation, but a subset of a far more pervasive ontological truth that pertains to objects of all types. The point here is not that we should exclude inquiry into human/object relations or social/object relations, but rather that these analyses are analyses for regional ontology, for a particular domain of being, not privileged grounds of ontology as such. The issue here is thus very subtle. It is not a question of excluding the human and the social, but of decentering them from the place of ontological privilege they currently enjoy within contemporary philosophy and theory. Nor does this entail that all objects relate to other objects in exactly the same way.[66]

Each chapter contains at least a short discussion of the machine's relationship with human beings and often the distinction between the human relationship and the object itself is theorised at length. But the human relationship with anthropomorphic technologies is almost always discussed in the narrowest of terms. The texts which form the evidence for these observations are a mixture of technical, philosophical and cultural works, providing material for the attempt to describe machine relationships with other nonhumans, including computers, floors, languages and concepts. Generally, exploring interactions between nonhumans provides greater insight into interactions between humans and nonhumans. We are limited, of course, by the possibilities of observation available to us. We can only witness relations between nonhumans that result in actions that we can sense and perceive.

Methodological questions take the theoretical foundations of this text out of its comfort zone. Creating a methodology that revolves around OOO is not easy. How can we learn about machines if we can never make direct contact with objects? This book merely presents some of the techniques that I have found useful to deploy OOO in aid of socio-political inquiry. This is a prevailing current in OOO, and a fortunate one. In 2018 Harman argued that OOO is of significant import in the age of "alternative facts":

> From a OOO perspective, there is no truth: not because nothing is real, but because reality is *so* real that any attempt to translate it into literal terms is doomed to failure. We can invoke *knowledge* against Trump's deceptions and evasions, but only insofar as we adopt a new definition of knowledge that incorporates elusive real qualities rather than directly masterable sensual ones [emphasis in original].[67]

In Harman's terms, we are capable of gaining knowledge about objects, even though the *truth* is beyond us. Consequently, there is a real object out there that has definite qualities, but because it is not directly accessible we risk being subject to billions of different human sensual relations with nonhumans. This is evident by the fact that a non-expert like Donald Trump can have a strong sensual relation with climate change through his denialism. In contemporary politics, the views of a non-expert are as impactful (or more so) as those of an expert. But by insisting that real objects exist, OOO resists the current post-truth environment. Greater knowledge of an object is possible through study. Expertise in climate change exists.

Harman's early work was somewhat apolitical, being more concerned with pure metaphysics. When a flat ontology is deployed there is a tendency to see each sensual relation as equally important and valid. But when it comes to politics, sensual relations must be grouped and assessed as more or less relevant. Metaphysics must be more interested in humans, and we cannot be contented with an ontology that dismisses expertise. But we are still left with the problem that no amount of knowledge of another object leads us to absolute truth. Hayler has described the possibility of an "asymptotic approach" to knowledge, one in which we may come closer and closer to a real object through "repeatable successful action".[68] Having studied the artificial agent SHRDLU, an object that few people outside AI history circles have heard of, I am confident that this book has greater knowledge of it than most objects do. The character of the sensual objects between SHRDLU and this book leads it to draw particular informed conclusions. So while the method is not scientific, it is one derived from the acquisition of knowledge, both by humans and by nonhumans.

Harman argues that a "theory of everything" is not to be found in the natural sciences.[69] He has several reasons for this, but one of the most important for the purposes of this book is that the natural sciences omit non-physical or incorporeal things. This means that things like concepts are left out, as are fictional objects. This is a significant weakness of the natural sciences, because concepts like gods and theories exert a tremendous influence on our world. This idea is returned to several times in the case studies, because since they are so closely connected to humans there tends to be a lot of influence of incorporeal objects like brands and political movements. We cannot easily observe these things through the natural sciences. The alien phenomenology of an anthropomorphic machine must pursue things in a transdisciplinary fashion to be most successful, with the deployment and problematisation of multiple kinds of expertise.

What does sensation mean for machines?

Machines experience the world in a very different way from humans. Their sensory organs mean that their relationship with the world may seem to be seriously impoverished compared with humans, although they may be capable of sensing their environments in ways that humans cannot. A personal computer senses information through a mouse and keyboard, which brings about certain kinds of experiences and reactions in the computer. But these are not the only germane sensory experiences possible. The computer senses and experiences data from a keyboard by making changes to its display. Perhaps it also senses and experiences the creation of a dent when it is accidentally dropped on the floor, albeit in a different way (its experience may be made evident to us by seeing it change its shape). The computer is in a relationship with its operating system, its programs, its network, the desk, the Earth (since it is fixed in place by the Earth's gravity) and the glass of water carelessly spilled on it. It is vital not to forget these other relations which contribute to the computer's rich experience of its world. In the near future, machines are likely to become significantly more agentic while remaining intrinsically reliant on alien sensation. The Internet of Things, a popular term describing the blossoming network of artefacts capable of exchanging data, exposes us to new deliberately engineered processes but also opens the possibility of all sorts of unthought-of relations between technological artefacts and other objects.[70] At this critical time, I believe it is important to develop new theories of relations between technological artefacts and other objects that attempt to understand the agency of the nonhuman. I have chosen to investigate sensation in order to emphasise not only the alien nature of the interface, but the alien nature of the technological artefact as a whole.

To this end, this book involves critical engagement with several key theorists and attempts to relate their work to anthropomorphic technological artefacts. It draws upon postphenomenology, Umwelt-theory and ANT, as well as OOO. In doing so, it will critique the ideas of several theorists in a close analysis of its case studies. It is thus both an empirical and a theoretical study. This book's point of departure from many of the theorists discussed in this book is that they write from an anthropocentric position, as has been the tendency for most of the history of the philosophy of technology, and indeed of philosophy more generally. Martin Heidegger, Jakob von Uexküll, Don Ihde and even Latour have seen the human perspective as a privileged position; the human is fundamentally different from all the other objects in the world, or it is the only perspective worth understanding. This book attempts a bidirectional study of humans and machines from a position that takes nonhuman agency seriously. But although this project owes a great deal to cybernetics, it is not simply mechanistic. Experience, sensation, mind – these are all words that will be critically explored.

This may sound like a lot of different theoretical approaches to address in one text. In *The Democracy of Objects*, Bryant describes himself as a "bricoleur, freely drawing from a variety of disciplines and thinkers whose works are not necessarily consistent with one another."[71] This book proceeds in a similar fashion, touching on multiple theoretical traditions that reflect the author's own intellectual background. But there is no intention here to provide an in-depth analysis of any of these approaches, only to discuss those aspects that pertain to the case studies and the decentring of anthropomorphic technologies. The main focus of this book is on OOO, and so to a certain extent all the other theory will be positioned through this frame. This is not part of an attempt to advocate OOO over other ideas, but rather because flat ontology permits a conceptual expansion in many diverse fields.

To what extent is the sensation of anthropomorphic machines relevant to their cultural role?

A machine's sensation and experience are highly relevant to its cultural, political and social position. It is evident that the way in which a machine senses its world affects its relationships with individual humans. When my Kinect detects my face, it begins a sequence of relations that end in me adjusting my behaviour to affect the phenomena visible to the Kinect. What is less clear is the way that sensation in machines affects their political and cultural positions. If the Kinect can identify me, then what questions of privacy and power are raised? Questions like this are anthropocentric but nevertheless significantly affect nonhumans. A machine may experience a serious change in its composition, relations and internal world if it is smashed with a blunt weapon. My Kinect would not sense me in the same way if it were unplugged and stored in a box inside a cupboard. In general, anthropomorphic machines are strongly affected by their relations with humans and with humanity.

The ethical issues associated with anthropomorphic machines are a frequently raised topic of public debate. These ethical issues are often connected to questions of machine sensation of other objects such as human personal data (in the case of the Kinect) and of pedestrians (in the case of self-driving cars). Natural language processing machines, natural user interfaces, and robots are all commonly connected to these kinds of issues because of their unknown and obscured sensory capabilities. This ties into larger questions about machine ethics. Can a technological artefact ever be morally responsible for its actions? Drew McDermott argues that it is not yet possible for a technological artefact to make ethical decisions because no technological artefact is yet capable of making a choice between "the right thing" and its own self-interest.[72] This is because, as McDermott says, no modern machine fulfils the criteria for really wanting anything. We can program a robot to be attracted to light but it is too easy to flip a switch and make the robot seek darkness instead. Without truly having desires (perhaps the desire to fulfil the needs and wants of its human user?) technological artefacts make decisions with moral consequences in the same way that they make any kind of decision. A computer may be programmed to start executing people when the Euro drops below a certain value, but it has not made a moral choice any different from if that computer were programmed to turn on a blinking light in response to the same stimulus. This book will argue that machines can and do have desires, in a thoroughly unhuman sense of the word. It is only by speculating about the internal world of an object that we can draw conclusions about what we might call its own ethical obligations. This is connected to a functionalist view of consciousness (see Chapter 3). The question of what it means to be conscious is highly relevant to the question of machine ethics. Kenneth Eimar Himma claims that an intentional state is necessary for agency, and without it no machine can make a moral choice.[73] There is a great deal of debate around machine ethics and the nature of consciousness.

Since alien phenomenology involves the study of nonhuman-nonhuman relations just as human-nonhuman relations, there must also be speculation about the phenomenal worlds of the nonhumans that surround the object of study. It is difficult to talk about the presence of an object in a room without describing the room itself. So numerous accounts are given of how nonhumans relate to one another. This is easier when we are speaking of physically embodied or corporeal objects, but it apparently creates difficulties when we begin to describe relations between objects and incorporeal objects like concepts or brands. But this is one of the most important ways that technological artefacts become culturally relevant, and cultural relevance is a huge part of being an anthropomorphic machine. One significant methodological distinction between this and other approaches to the philosophy of technology, and even where it deviates from some other applications of OOO, is that it emphasises concepts as objects. One of the prevailing themes in this book is its struggle and attempts to differentiate between different objects that may be grouped together by humans under the same word. Anthropomorphic machines are such potent ideas that the concept of a machine may be radically different from the machine itself, arbitrarily grouped by humans using unhelpful language that exacerbates problems with machine ontology and, more practically, encourages hype and misleading statements. The concept of a machine obeys the laws of concepts, and its existence in the world and relations with the world are entirely different from the machines themselves. Both (or all) objects that we thus group together commonly have significant bearing on one another. For example, the idea of sexbots is a significant motivator for the way that humans approach specific sexbots. The specific sexbot also has a relationship with the concept of sexbots, the nature of which is discussed in Chapter 5. In Chapter 2 and 4 there is yet another complication: incorporeal objects (games, programs) that themselves are instances of another incorporeal object (an idea or brand). Take for example the book that you are reading now. You might have a hard copy in your hands or an electronic copy on your screen. But while that specific book is its own object, the book also exists as an object that transcends physical instances. *Harry Potter and the Prisoner of Azkaban* is an object; my copy of *Harry Potter and the Prisoner of Azkaban* is an object. Unfortunately, the two objects have the same name so confusion may occur. The same applies to the objects of study in this book. They exist in specific incarnations, but often also as ideas that are not limited to platform, location, or material composition.

Structure of this book

The four case studies in this book are described in a very rough ascending order of their apparent resemblance to human beings. They are, however, described in different theoretical terms and are used to reflect upon the writings of different theorists relevant to the nonhuman turn. It is hoped that some evolution of ideas is evident through this book, but each case study is relatively self-contained. And this is as it should be. Each case study is as radically different from the others as they are from humans with respect to their modes of existing in and navigating the world. The superficial taxonomy of technological objects typical of human descriptions is not all that relevant to their internal experiences. Certain similarities are observed, but the differences between them are fundamental. Some texts associated with the nonhuman turn are deliberately constructed with this in mind. For example, *Object Oriented Environs*, a collection of essays arranged by the authors' names, is intended to be read in an order determined by "some object that you will allow to exert its aleatory agency over your reading" such as a twenty-sided die.[74] It is best to read the chapters of this book in the order they are presented, but while bearing in mind the independence of each case study. Regardless, Chapter 1 should be read first for the background on OOO that frames the studies of technological objects in Chapters 2-5.

The Microsoft Kinect (Chapter 2) is an add-on to the Xbox line of videogames consoles, comprising of a sophisticated camera to detect human movement and a microphone to detect voice commands. When first released, much of the discussion about it centred on its ability (or inability) to "see" and "hear". The Kinect for Xbox One is the first case study for two reasons. Firstly, it is comparatively (or deceptively) simple when compared with the other case studies. There is no visible movement in the Kinect. It does not appear to have a great capacity to communicate its experience to human beings. Secondly, because the Kinect is almost universally portrayed as a watcher and listener (both in Microsoft's promotional material and by users – see Chapter 2), it is an ideal case study for beginning to explore this book's guiding questions. What does sensation and experience mean in the Kinect, and what can we say about it?

Much of the research for Chapter 2 was conducted in the first half of 2014, shortly after the release of the Kinect for Xbox One and a time when the evidence of human first-hand experience of the Kinect's sensory and perceptive systems was ample. User accounts were gathered from online forums describing a sort of collective effort to engage with the phenomenology of the Kinect. Individual users described their attempts to coax the Kinect into seeing and hearing them correctly. This is captured in this book with two examples: the "Xbox on" problem and the videogame *Kinect Sports Rivals*.[75] The Kinect often proved unreliable. Difficulties in persuading the sensor bar to detect human bodies and faces have resulted in the alienation of certain sectors of the "gamer community", notably the self-described "hardcore" gamers. It has proved difficult to persuade these users to form relations with the Kinect, which has had an effect on the cultural relevance and reputation of the device.

An essential part of this analysis is the distinction between the different types of object that are erroneously conflated by the signifier "Kinect". The Kinect is many things to different humans: a harbinger of more casual games for the Xbox One; an unnecessary but compulsory add-on to an expensive console; a practical voice-activation technology enabling contactless direction of the Xbox One; a means of engaging in physically energetic play; the subject of online trouble-shooting; and, a potentially malevolent eye spying on your living room. But it is also an individual Kinect, beneath my television and connected by a cable to my Xbox One. The term "Kinect" becomes insufficient when asked to represent two such different levels of identification, so this chapter employs a simple method of distinguishing between the brand and the situated artefact. This book contends that these are two different objects, not two different aspects of a single object.

This apparent bifurcation of physical and cultural identity is also present in William Grey Walter's robot tortoises, the so-called *Machina speculatrix*. Chapter 3 builds on the idea of this division within a human-made object and goes into detail about the tortoises' sensory world. Built in the late 1940s, the robots existed as sensing and moving agents but the human interest in them is more strongly connected to what they indicated about living things. Walter was a cyberneticist, and in his view the robots confirmed his ideas about life, sensation and identity. Jussi Parikka wrote about the tortoises in his book *Insect Media* and briefly discussed them in relation to the ideas of Jakob von Uexküll.[76] Chapter 3 continues this analysis and extends it by drawing on contemporary panpsychism. There are, then, two steps to this process: the acknowledgement of the tortoises' sensory worlds followed by an appreciation of the heterogeneity of nonhuman "minds". The two theoretical structures are occasionally in conflict with one another, and this is also explored.

The tortoises, which were entirely synthetic, were anthropomorphised in many ways. They possessed eye-like structures capable of sensing light and contact switches that enabled them something like a sense of touch. In this way, they could find their way back to their hutches when it was time to recharge their batteries. Walter gave them names and studied them as an ethologist might plot the progress of an ant or mouse. Possibly the ease with which these simple machines can be anthropomorphised promotes the application of panpsychist ideas. In other words, the blurring of the line between human and nonhuman suggests that the mind is not exclusively human, but is in some way a universal quality of matter. However, it is important to avoid prioritising minds in this analysis, since that could mean the inflation of a human idea beyond its importance in a nonhuman object. The mind must be treated like any other object. Consequently this book questions whether sensation and experience are best interpreted as metaphors in nonhumans or as the product of a conscious mind, but the chapter ultimately concludes that this is unknowable and irrelevant to the study of nonhumans by humans.

Chapter 4 is less focussed on the internal world of nonhumans and more interested in the ways in which objects extend the sensation and perception of other objects. The case study presented is the natural language processing (NLP) program SHRDLU, built by Terry Winograd between 1968 and 1972, and in particular its relations with the human user (and vice versa). The concept of postphenomenology, particularly that of Don Ihde, is placed in a more anthrodecentric context for this purpose. Postphenomenology is an example of a phenomenological approach that places greater emphasis on the role of nonhumans than traditional phenomenology, however it is still too one-sided to fit entirely comfortably into this book.

SHRDLU is a language-user and possesses the capacity for sensation and action within its own tightly constrained world, a world in which blocks of different colours and shapes are stacked and restacked by a virtual robot arm. It is designed to understand a human user's commands, and to answer questions about the way the blocks are stacked, and about its own actions. Because of its extremely limited world and capacity for action, it is an excellent model for elementary philosophical inquiry. Like the robot tortoises, which are all thought to be destroyed or otherwise removed from public consumption, it was not possible to interrogate Winograd's SHRDLU directly. The analysis relies on first-hand accounts and on a version of the software remediated for an early version of Windows.[77] Nevertheless, with these sources it is still possible to interpret the way in which SHRDLU and the human extend each other's worlds.

Of all the case studies presented in this book, the sexbots and gynoids of Chapter 5 are the most human-like in their physical appearance. They require physical humanlike form even if their sexual organs are the only ones required for the act of coitus, or if their faces are the only ones required for verbal communication. They are also subject to significant political and cultural scrutiny. Sexbots, of course, can be of any or no gender, but the focus here is on gynoids – that is, feminine-presenting humanoid robots that, in theory, are to androids what women are to men. Some space is devoted to the cultural role of the gynoid, which is highly dependent upon the metaphysical commitments guiding their development and presence in different cultural contexts.

The real focus of the chapter is on the *concept* of gynoid sexbots and how we can study a concept through the interrogations of other objects (such as texts, humans, real specific gynoid sexbots, etc.). A great deal has been written about gynoids, both in popular fiction and in academic contexts. However, very little interest has been shown regarding their experience of the world. The sensory capabilities of a gynoid reveal a great deal about the concept of gynoid sexbots, as they tend to be deliberately engineered into them by humans. Those robots exist in a bidirectional relationship with the concept. It is hoped that this chapter on gynoids might demonstrate how OOO can be a powerful tool for stimulating political discourse, in response to criticism of its apolitical nature. There is also some criticism of OOO and suggestions of how alien phenomenology might best be deployed. As in previous chapters, the focus is twofold: an in-depth analysis of the case study, plus the acknowledgement that ontology guides the cultural criticism of technological artefacts.

Chapter 1
ANT and OOO for technology

Different writers have approached the question of technology within different prevailing paradigms, often as an extension to other philosophical or methodological commitments.[78] There is significant variation in the estimation of how much technology can be said to affect human society and culture, and how much cultural forces shape technological artefacts. Over time, machines have come to occupy a vast spectrum of complexity, making it difficult to encapsulate them with any single political or cultural theory. There is no exception to this for OOO. OOO tries to describe the reality of objects and their relations with the rest of the world, but that doesn't in itself make it a good tool for analysing anything. Even though it is an increasingly useful tool for discussing technology, it is still just one tool out of many. But it prompts an unusual approach to talking about machines, one that I think brings certain things to the fore of the analysis. For one thing, OOO has a great power to alienate the human and distract from the "user", which is a great asset in the study of anthropomorphic machines.

This chapter will briefly present the key insights of OOO as metaphysics, which was the foremost emphasis of Harman's early work. Hopefully those who are new to OOO will find it a good explanation, but the main aim of the chapter is to show the relevance of OOO to studies of technology. The applicability of OOO to disciplines other than philosophy is the focus of more recent OOO literature. This chapter will turn to how OOO can be turned towards the study of specific objects, and especially objects with which humans are strongly concerned. The transgression of disciplinary boundaries is essential to this. There is a gap between science and technology studies (STS) and metaphysics because one is implicitly concerned with specific instances of human-nonhuman relations and the other is implicitly concerned with *everything*.

Before talking about OOO this chapter will talk about actor-network theory (ANT). OOO is by no means a successor to ANT, although they are sometimes compared. They are different tools with different implications. Harman's study of ANT has led him to conclude that while OOO is led by objects in themselves, ANT reduces objects to their mutual effects on one another. While OOO's objects are fairly passive, ANT's actors or actants are very active. Actants also have reciprocal and symmetrical relations, whereas in OOO relations are asymmetrical and may be non-reciprocal.[79] ANT also has origins in the study of science and technology and is very relevant for contemporary STS. By contrast OOO has its origins in phenomenology and is therefore a roundabout route to the study of technology. Yet there is a similar cadence to the two fields that may be apparent to the reader.

Actor-Network Theory (ANT)

Bruno Latour was central to the development of ANT as part of his studies of science and laboratory practice. ANT is self-consciously non-anthropocentric and was developed by Latour and others such as Michel Callon and John Law to explain the work of scientists and engineers in interaction with the world. A particular hallmark of this strategy is the commitment to considering human and nonhuman things in a flat way. As Harman has noted, ANT scholars are therefore different from most other post-Kantian Western philosophers who write within the paradigm of "the bland default metaphysics that reduces objects to our human access to them."[80] Each human or nonhuman actant has its own agency and is constantly changing its relations with other actants. No actant is ever reducible to any other actants, as this would mean creating a hierarchy of objects. For example, a human body is not reducible to organs, systems, cells, evolutionary history or demographics. Each of these actants is identifiable as an entity but are in relation with the other actants.

The agency of nonhumans is revealed by ANT to be key to the success or failure of scientific projects. For a program to succeed, the scientists must engage with all the relevant human and nonhuman actants and persuade them to behave in a particular way. A classic example is Callon's study of St Brieuc Bay in which marine biologists attempted to find out how to increase the number of scallops living in the area.[81] Callon demonstrated that in order to make the project work, the scientists had to attract the local authorities and laypeople and persuade them not to sabotage the study. But he also showed that the scallops themselves were active participants in the network and needed to be encouraged to behave a certain way so that the scientists could study them. The human or nonhuman, and even the living or non-living status of the actant is only relevant insofar as it affects its agency. Agency and ontology are generated through relations between objects. In Latour's ANT, objects only really exist as a set of relations, shifting ontologically and mutating constantly in response to the entities around them. A leaf sitting in the palm of my hand is constantly embroiled in competing networks. The wind wants to push air particles through the space that the leaf occupies, moving it gradually out of the way. The air wants to drag water molecules out of the leaf's stoma. The leaf is also part of my network, behaving in a typical and uncomplicated way, allowing me to discuss its metaphysics. The leaf is subject to plants, humans, ideas, physical forces, subatomic particles, radiation, and time, constantly changing as each of these objects moves around it. The leaf's agency and ontology emerges from the sum total of these networks.

Since the 1980s, ANT has been used widely beyond science studies.[82] Yet it may be said that even ANT is too anthropocentric. Latour's ideas have been reinterpreted by Harman as metaphysics in his book *Prince of Networks*.[83] Harman is critical of ANT's relational ontology and the way Latour construes the agency of objects, but praises the "power and precision"[84] of his ANT. Speculative realism has been influenced and critiqued by scholars from many different backgrounds, including (but not limited to) Latour.[85] As mentioned in the introduction, speculative realists are united in their rejection of correlationism in philosophy. Correlationism is the assumption that all human research, analysis and endeavour is done for the ultimate benefit of humans. In the words of Quentin Meillassoux, correlationism insists that it is not possible "to consider the realms of subjectivity and objectivity independently of one another."[86] Anti-correlationism does not distinguish between the existence of an object and the way that a human encounters an object; an object is not the correlate of human thinking. In the words of Levi Bryant, "we must avoid, at all costs, the thesis that objects *are* what our access to objects *gives* us" [emphasis in original].[87] Correlationism carries the implicit belief that humans are special and unique, perhaps even that we are a special ontological case, and is based upon the increasingly problematic binary of "human" and "nonhuman". ANT's emphasis is mostly on the relations between objects. The objects themselves only exist in the background of processes of translation and delegation. Ian Bogost claims that in ANT "entities are de-emphasized in favor of their couplings and decouplings. Alliances take center stage, and things move to the wings."[88]

The context and emphasis of disciplines is a not insignificant indicator of their metaphysical commitments. ANT is a sociological methodology, and therefore must privilege human priorities above all others. In Callon's study, he only discusses the agency of the scallops and other nonhumans with reference to how human actants may manipulate them and arrange them into networks. The agency of nonhumans is something that must be *overcome*. ANT is a part of the epistemological project in the philosophy of science; scholars work to understand how science progresses and how humans gain information about the world. Harman, however, insists upon maintaining a distinction between the ontology of objects and epistemology.[89] These fields may consider objects independent of all other external actants, including humans, and the nonhumans are therefore better represented. Harman places Latour's work in an ontological context to give his own work greater depth and a sense of legacy, but Latour himself has typically been focussed on the history of science in an effort to understand the success or failure of particular human endeavours, such as the sequencing of a peptide or the attempt to implement a new public transport system in Paris.[90] Harman and others, in contrast, focus on how individual objects exist in the world and how they may be said to interact. These ideas have implications for specific episodes in STS, but they can also exist as purely theoretical disciplines.

Object-Oriented Ontology (OOO)

Speculative realism has spawned several different anti-correlationist schools of thought, and OOO is one of the most well-known. OOO is an emerging field, and it is prone to contributions by individuals from many different disciplines. Harman initially used the phrase "object-oriented philosophy"[91] (but has latterly embraced the term OOO, as is evident in his use of the term in a recent book title[92]), Levi Bryant invented the word "onticology"[93] to describe his particular approach, and Ian Bogost prefers to think in terms of "unit operations"[94] to distinguish his ontology from object-oriented programming. OOO seeks to reverse the linguistic turn in Western philosophy and social science and return them to a study of materials. It is interested in objects for themselves, not objects for humans. The word "object" is used here in an extremely broad sense. It does not only refer to the simple inanimate objects accessible to humans. According to Harman, objects do not need "to be physical, solid, simple, inanimate, or durable...".[95] Tom Cruise, Superman, orange blossom, *Drosophila* flies, manganese atoms, pulsars and the concept of free will are all examples of objects. Like in Latourian metaphysics, objects are not arranged into hierarchies, nor are they reducible to more fundamental explanations. Timothy Morton argues that OOO has arisen in the wake of the threat of environmental apocalypse.[96] It rejects "Nature" as a concept imbued with correlationism, and it also "offers a middle path – not a compromise, but a genuine way out of the recent philosophical impasse of essentialism versus nihilism."[97] In the modern world, humans are becoming more aware of the agency of powerful nonhuman actants, such as climates, carbon footprints and oceans. All these objects seem to exert an agency independent of humans, particularly individual humans or small groups. They are objects without subjects, and thus lend themselves to a non-correlationist approach. OOO studies all actants as objects, not as what happens between subject and object.

Harman is arguably the best known object-oriented ontologist. Since the publication of *Tool-Being* in 2002 he has since written several books and papers on OOO. In *Tool-Being*, Harman closely analyses the work of Heidegger, whom he clearly admires, although his interpretation is unusual.[98] Timothy Morton states that Harman:

> ...discovered a gigantic coral reef of sparkling things beneath the Heideggerian U-boat. The U-boat was already travelling at a profound ontological depth, and any serious attempt to break through in philosophy must traverse these depths, or risk being stuck in the cupcake aisle of the ontological supermarket.[99]

The first of Harman's major discoveries comes through the study of Heidegger's tool analysis in *Being and Time*. Heidegger distinguished two types of relationships between humans and machines, expressed through his well-known example of the use of a hammer. The hammer is generally ready-to-hand (*zuhanden*), available to us as a tool and untheorised. Harman uses "tool-being" as a synonym for readiness-to-hand.[100] Tool-being is apparent when the tool is used unconsciously to achieve a goal. Very rarely, we might need to theorise the hammer, such as when it is broken.[101] At such a moment, the hammer is said to be present-at-hand (*vorhanden*). The "broken tool" is a metaphor for when objects become noticed, theorised or even just gazed at. Most of the time, objects are ready-to-hand for humans. Objects appear in the background of our lives, taken for granted, and are not the centre of our attention. Objects that work well from a human point of view become invisible. For instance, most of the time a screen is taken for granted by us, but it becomes present-at-hand when we notice that it has started to malfunction, or when we begin to theorise it. Harman makes an implicitly realist claim about objects: that the tool-being of objects is utterly separate from the way that the object appears to other objects.[102] Harman claims that the multi-faceted configurations of objects are incommensurable with each other since the readiness-to-hand (or "tool-being") of an object always rushes ahead of its presence-at-hand.[103] Each interaction with an object is different, and no other object can ever encounter the whole of an object's tool-being. Because of the infinite number of objects in the universe, there are therefore an infinite number of different present-at-hand experiences, each of which is invisible to other objects. From this, Harman concludes that objects withdraw infinitely from all relations. An object can only be ready-to-hand or present-at-hand at one moment, and presence-at-hand reveals only those qualities that we are equipped, at that moment, to sense.[104] So no object can experience *all* of another object, or gain access to the object's true essence, its inwardness.

Harman's retelling of the tool-analysis means that readiness-to-hand and presence-at-hand are always at play in interactions between all objects, including inanimate ones. Harman's description of tools originates from Heidegger's famous account of hammers, but the principles described by Harman need not only apply to "tools" in the conventional sense. The model developed by Harman applies to every object, including both human and nonhuman objects. Readiness-to-hand and presence-at-hand are types of relations between two objects, in this case the human and the hammer. Yet readiness-to-hand and presence-at-hand are, for Harman, evident in every kind of object relation. This is not simply a panpsychist argument, although panpsychism or panexperientialism have been embraced by certain participants in the nonhuman turn (see Chapter 3). Harman himself asserts that Heidegger's concept of Dasein, a word that describes human being, has been incorrectly interpreted to refer only to humans. Harman argues that Heidegger spoke of human Dasein but that he did not necessarily intend to restrict Dasein to humans.[105] In Harman's view, the kind of being that Heidegger spoke of referred to any being that was concerned with its own being, which was not necessarily limited to humans, although Heidegger probably would not extend Dasein to inanimate objects.[106] However, the extension of Dasein to all objects is not necessary for Harman. For him, readiness-to-hand and presence-at-hand become radically separated.[107] He identifies a depth in an object's readiness-to-hand that cannot be fully experienced by *any* object, not just a human. He gives the example of a metal appliance resting on a lake of ice:

> I hold that the resulting interaction between stove and ice is philosophically *identical* with the more familiar case of Dasein and the broken hammer. For what is decisive in the famous account of the "broken tool" is not that implicit reality comes into conscious view, as if *human surprise* were the key to the reversal within being. Rather, the important factor is that the heavy object, while resting on the ice as a reliable support, did not exhaust the reality of that ice.[108]

Presence-at-hand is not something that only happens when a human interacts with an object. A stove only reacts to certain things about the ice: hardness, coldness, brittleness, and perhaps other things. The hardness, coldness and brittleness of the ice do not constitute the whole of its being. Coldness is a "relational property", and in Harman's later terminology constitute a part of the sensual object formed between stove and ice.

Unlike in ANT, Harman's objects do not become different objects when exposed to relations with other objects. They are withdrawn from the rest of the universe and "interact only by way of abstracting from each other."[109] Objects are separated from other objects' access to them. A human might experience an apple in terms of redness or, at another moment, in terms of ripeness. But these are only a human's conception of an infinitely withdrawn object. The apple-ness of the apple is inaccessible. The apple may also have an experience of the human, though this kind of bidirectional relationship is always asymmetrical.[110] Two separate interactions occur. For example, the human may experience the texture and taste of the apple as they bite into it, while the apple experiences the piercing hardness of the teeth by having a section removed. Two metaphysical processes occur at the same moment. Two separate interactions also occur when two nonhuman objects relate to one another, but we must speculate about the senses of nonhumans. If a stone sinks into mud, then both the stone and the mud encounter certain qualities of the other. We might speculate that the stone senses a downward movement and a wetness that adheres to its surface. The mud changes shape, sensing the weight of the stone and reacting to it. Perhaps each of the nonhumans experiences a slight change in temperature. But the mud does not encounter what we would call the *colour* of the stone, because it does not have the faculties to sense light (to my knowledge). Note that this does not imply that objects are completely immutable for Harman, only that the object does not become another object moment by moment. The apple, for example, changes when it is bitten, and its real qualities change correspondingly. Once the apple is eaten or decayed in a rubbish dump somewhere, it is destroyed and ceases to exist. This is dealt with further in Chapter 3 in the discussion of the withdrawal and change of robots.

No two objects can ever make direct contact. Harman speaks of a "mutual darkness"[111] of objects, a withdrawal that requires "a third term or mediator"[112] for contact to occur between the "noncommunicating crystalline spheres sleeping away in private vacuums."[113] The world of real objects needs something to explain interaction. Here Harman takes inspiration from José Ortega y Gasset's distinction between real objects and the access that humans have to reality through metaphor.[114] The influence of Ortega is evident in Harman's frequent use of metaphor. Ortega distinguishes between the "images" of objects, which is what we encounter when we look at or use an object, and the reality of objects.[115] Objects exist in their own right, and Harman calls them "real objects". Ortega's "images", which are "nothing more than correlates of our own experience," are "sensual objects".[116] Real objects *really* exist – that is, they have their own reality outside of our perception of them. The sensual object is analogous to Husserl's intentional objects; that is, the sensual object is the one that exists to another object. But it is not "an idealist prison" removed from the inaccessible real world, as in Plato.[117] It needs to have a corresponding real object that other objects can only access indirectly.

Real objects exist for themselves and still exist when we turn away from them. Sensual objects exist only to facilitate relations between real objects permitting indirect interaction between the two withdrawn objects, so that "the actors involved in [the interaction] are no longer separate, but form a new object with its own interior."[118] In the example of the apple, the human does not experience the real object but aspects of a sensual object such as "sweetness", "redness" or "hardness". The sensual object is a mediator between the apple and the human. It is not possible, for Harman, to claim that the human is interacting with a partial object, namely, that part of the apple that is sweet. He rejects Hume's characterisation of objects as "bundles of qualities".[119] Objects are inherently irreducible and whole. Therefore, the apple is withdrawn and inaccessible to every other object, no matter how many interactions occur. This indirect interaction between objects themselves is what Harman calls "vicarious causation".[120]

The myth of the off-shore drilling rig from his book *Circus Philosophicus* may help to explain this position.[121] Harman compares the lack of direct interaction between his objects to the occasionalism of medieval Islamic philosophers, who claimed that objects could only interact with God.[122] In this view the human and the apple do not interact directly: each object is in a relationship with God, and thereby indirectly affect each other through divine intervention. In other words, God acts at every occasion. But in Harman's ontology each object is reduced to an image that is conveyed by a sensual object.[123] God is replaced by the metaphor of the oil rig, and each object is reduced to an image. (Harman appears to borrow "image" from Ortega as a substitute for "sensual object".[124]) We can only obtain an image of each of these objects, not the objects themselves; that is to say that the metaphorical oil rig receives metaphorical sensual objects but not real ones. Every object becomes "a hazy caricature of its deeper plenitude"[125] to the other objects. Oil rigs, in Harman's poetic description, convert any number of dead organic objects into fuel by siphoning them out of the earth: "It draws them to the surface of everyday life, where they are used as energy for the most prosaic modern actions."[126] Before siphoning, they are decayed accumulations of hydrocarbons from the time of the dinosaurs. After siphoning they are crude oil, and become tangled in a set of human concerns and ideals. The oil rig turns past objects into present objects.

In the myth, it is not only objects from the past that are transformed, but objects from the present and the future; living, dead and non-living objects; human objects and nonhuman objects. In the above example, the oil rig is the human and the image is the sensual object that comes from the apple. Harman envisions each object in the universe as an oil rig, equivalent in their inability to access real objects through interaction, all siphoning images of all other oil rigs, so that "[a]ll real objects of every size now have the power to interact with all other things, at the price of turning them into images."[127] The real objects do not touch each other directly, but can interact through sensual objects. Siphoning does not reduce the object. The real object remains withdrawn, and only images of it are accessible through the oil rigs. Harman offers this myth as a metaphor for vicarious causation. In this metaphor, it is also impossible to represent the materialist view of symmetrical interaction, since oil rig A cannot interact directly with oil rig B. The apple siphons an image of the teeth while the human siphons an image of the apple. Harman's description is a sublime account of object relations, and it succinctly argues for the lack of legitimacy in any anthropocentric metaphysics. Nevertheless, it is a challenging position to accept.

Over time, Harman has become more and more interested in art and aesthetics, even declaring that aesthetics is "first philosophy".[128] This is based on his concept of "allure", which he describes as the disconnect between the real object and the sensual object. (This disconnect is discussed further in Chapter 3 with reference to Edmund Husserl.) The disconnect between real object and sensual object is evident in human uses of metaphor in the description of objects. Harman gives the example "my heart is like a furnace".[129] It is a metaphor that evokes certain furnace-like qualities (such as intensity or warmth) while explicitly ignoring others (such as size and metallic composition). In the case of the metaphor, then, it is the spectator or reader who is the "real object" to which sensual qualities bind.[130] Harman's metaphysics makes a point about aesthetics that exposes one of his most important concepts. It represents in very clear terms the withdrawal of objects. It expresses the need for a metaphysical explanation of the way that art can "move" humans without physically touching us; as Francis Halsall points out, this is appealing to artists because it supports the idea "that works of art have an autonomous identity"[131] (although, like all objects, they are only as autonomous as their relations with others allow). Emmy Mikelson makes a similar point, commenting on the artwork as an object in its own right to which observers only have a single point of entry.[132] Mikelson argues that speculative realism has "a constructive rapport" with the beheading of the subject in art.[133] It is curious that such an anthrodecentric approach should find such use for something as human as art, and in a way it exposes the paradox inherent to attempts to find a non-correlationist metaphysics.

Ian Bogost's metaphorism draws on this part of Harman's metaphysics. The effect of one object on another can be depicted as a process of one thing becoming like another, either physically or figuratively.[134] When I place a mug on a table, the mug models itself on particular qualities of the table: immovability and position in space. As a plant absorbs water through its roots, the water conforms to the shape of the tree, adhering to the walls of the xylem and shooting up towards the leaves. As a text is read by scholars of a different tradition, it is translated into a different technical language and forms analogies with ideas that the author did not intend.[135] This resembles the myth of the oil rigs because each object detects a caricature of all others, but without requiring us to accept the concept of vicarious causation. Of the two accounts, I prefer Bogost's, mainly because it seems less reliant on a set of *a priori* assumptions derived from yet more ancient philosophy. OOO is preoccupied with new approaches that radically differ from almost all older ontological models. It therefore seems incongruous to articulate a complex theory of object interaction that has its origins in medieval Islamic theology. I am not alone in this discomfort. Levi Bryant has said that "[p]erhaps no element of Graham's thought has been more maligned than his doctrine of vicarious causation."[136] Harman, however, brings OOO more into line with traditional phenomenology and therefore lends the discipline a sense of authority that it might otherwise lack.

Alien phenomenology

We might conclude that OOO is in conflict with phenomenology. Traditionally, phenomenology claims that the human experience of objects is the only possible way of knowing the world. The existence of objects independently is of little interest and, in some cases, impossible. Tom Sparrow has explored the contradiction in phenomenology.[137] Sparrow argues that while phenomenology retains a belief in the reality of objects, phenomenological research would in fact do better with a commitment to idealism rather than realism.[138] When compared with speculative realism, its weaknesses are revealed: it can only study how conscious entities relate to the world.[139] But in OOO, and other speculative realisms, the implicit realism is evident. Objects really exist and two non-conscious objects can form relations. OOO claims that objects have inherent qualities that do not emerge from relations. This is particularly clear in Bogost's work: phenomenological ideas may simply be extended to apply to inanimate objects as well as humans. Instead of studying how objects appear to us, we study how objects appear to nonhumans. His book *Alien Phenomenology, or, What It's Like to be a Thing* asks what sorts of phenomenological experiences machines (and other objects) are capable of, without making hierarchical distinctions between classes of objects.[140] Levi Bryant reframes Bogost's argument in terms of his own ontological commitments by suggesting that alien phenomenology seeks "to determine the flows to which a machine is open, as well as the way that machine operates on these flows as they pass through the machine" (Bryant is not only referring to technological artefacts when he speaks of "machines").[141] Alien phenomenology, therefore, involves making inferences about objects and the kinds of interactions that they have with other objects, based on observation of the structure of the object and the kind of relations that it forms. It has been called "*applied speculative realism*" [emphasis in original].[142] Being based on speculation, alien phenomenologists can only make second-hand observations about nonhumans. Metaphysically, this is a problem. But this is a strength as well as a weakness.[143] In the case studies in later chapters, it will be shown that anthropomorphism in technological objects is a persuasive tool, and it is argued that alien phenomenology presents a means of arriving at a more critical study of objects.

There are now numerous scholars devoting time to applying anthrodecentric ideas to nonhumans. The Parliament of Things project publishes submissions from authors who write in a Latourian style about the existence of nonhumans, as well as identifying this style in earlier authors.[144] Also, as mentioned in the introduction, Bogost and Christopher Schaberg have edited a series of short books intended to provide insight into the being and relations of objects. This series encompasses many mundane objects. *Eye Chart* by William Germano describes the historical background to contemporary eye testing equipment, and *Drone* by Adam Rothstein explains the relative significance of different engineering traditions, as well as the software and hardware involved in contemporary drones.[145] The series exposes un-interrogated relations between physical objects (especially human-made ones) and seemingly unconnected ideas, texts, and physical objects (especially historical ones). This, in my view, is an important part of alien phenomenology, although not all of it. Often, the work of alien phenomenology of the object is left up to the reader in a way that erases the background and theoretical connections of the author. John Garrison's *Glass* is a good example of this.[146] The title is simple and broad enough to highlight the various ways that the word "glass" is used by humans, and the themes that bind them together: "the matrix of time, self-reflection, desire, and world-creation that this book has been tying together – and tying to our cultural fascination with glass."[147] This fascinating history connects mirrors, microscopes, sea glass, and Google Glass (the Augmented Reality artefact).

But the broadness of the word "glass" is also a limitation, because the tying together of glassy objects from so many temporal and physical contexts is a very human practice. The objects are connected through language and analogy, but *from the point of view of the sea glass*, Google Glass is a very alien object. For us, the analogy becomes reality, and the qualities that set these objects apart are less emphasised than the qualities that keep them together in our minds. For this reason, Garrison's account is anthropocentric. Garrison's focus is the *concept* of glass, rather than any of the corporeal objects they describe. It is a tool that the alien phenomenologist can use, certainly, and it is used frequently in this book. We sometimes group together vastly different phenomena under a heading that make sense to humans, not machines. In this case, grouping objects together under a heading like this enables the human practice of alien phenomenology which cultivates interest in their way of being in the world. In this way we can expose the arbitrary nature of those anthropocentric groupings that feel natural to us, or that we have perhaps inherited from other humans.

Knowing more about the relations between sea glass and other objects is useful for discovering its inner qualities. But it can only provide a partial glimpse into the phenomenology of a glassy object. This is not a criticism – no author can provide a complete account of sea glass, since we are not the sea glass. One book in the series that goes some way to evading this issue is *Earth* by Jeffrey Jerome Cohen and Linda T. Elkins-Tanton.[148] Co-writing a book on this subject, particularly by scholars in different fields (Cohen directs a medieval and early modern studies institute and Elkins-Tanton is a planetary scientist), brings the partiality of human attempts at alien phenomenology into the foreground. Parts of the book are taken from conversations between the two scholars, and their disciplinary training is evident in the way that they speak about their area of study. As Cohen says:

> Earth is an object of such immensity that it beckons us to think outside of our comfortable orbits of career path and discipline. The risks must also include that we would talk at each other (your Earth might be a planet with a history so long that humans simply do not figure much; mine might be an Earth that is too much a home for humans and does

not adequately engage with the longevity of stones, water, or atmospheric gases).[149]

The two authors are still communicating in a human way and are inevitably drawn to anthropocentric inquiries. But their cross-disciplinary collaboration forces them to confront the truth that both their views are partial; both their views are only part of the story for Earth. Earth itself has no way of providing humans with a complete account of its being and its relations with other objects. But there is no reason why we cannot attempt to know as much as possible and thus arrive at a more detailed and faithful alien phenomenology. This is closer to the sort of methodology that we could use to apply OOO to the study or interpretation of specific objects.

Human interpretation of phenomena is less likely to be questioned than machinic interpretation of phenomena. We are more likely to believe a human view of events. This is despite evidence of humans' underwhelming ability to detect what is "really there" and our tendency to be fooled by illusions. For centuries scientists have become more and more aware of the subjective nature of human vision and we now know that the mind is not a passive receiver of information but is actively involved in sensing and perceiving.[150] Eyes are prone to error and eccentricities. But they are *human* errors and eccentricities. Light is received by the eyes and is immediately a part of the complex biosemiotics of the human eye involving reactions of photosensitive material, changes in ion concentration and the release of neurotransmitters. No inorganic material is capable of carrying out this sequence of events, and the sensed and transmitted information conveyed by machines is subject to completely different encoding and decoding. We perform similar sensory-perceptive acts in our relations with all objects, including concepts and other incorporeal objects. Although every object (be they human, machine, or idea) has a sense of the world, humans are typically better at conveying information to other humans. It may be a challenge to describe what I see, but because most humans share the same visual apparatus other humans are likely to grasp my take on events. If a machine were to describe its experience of the sensation of light and colour, it might not be so easy to understand because artificial sensors are not structurally similar to human eyes. Similarly, I am more likely to understand Cohen and Elkins-Tanton's take on Earth than that of the Earth itself, since I am not attuned to magnetic fields, gravity, and tectonic shifts that span long beyond my own lifespan. Alien phenomenology will always be anthropocentric to some extent.

Anthrodecentric thinking has yielded many accounts of ordinary material. Jane Bennett's vital materialism is distinct from alien phenomenology, and it is heavily informed by the tradition of Deleuze and Guattari,[151] but her description of "*Thing-Power*: the curious ability of inanimate things to animate or act, to produce effects dramatic and subtle" [emphasis in original][152] suggests a similar anthrodecentric goal to that of alien phenomenology. This is particularly evident in her analysis of metal and metallurgy.[153] Bennett describes the many ways in which metal has traditionally been portrayed as passive, dead, and uniform.[154] However, she points to both contemporary scientific and ancient metallurgical texts that highlight the activity and idiosyncrasy of individual metallic objects. Some of our reluctance to appreciate this activity is due to human limitations, such as the shortness of our lifetimes and inability to perceive the tiny imperfections, holes, and quivering of atoms that give metallic objects their unique properties.[155] Similarly, in *Stone*, Jeffrey Jerome Cohen describes the nature of stone (in everything from cultural to tectonic terms).[156] Cohen's background in medieval studies naturally leads to the work being concerned with the connotations of stone within that time and place, and like Bennett's work it is peppered with cultural references. It is an anthropocentric approach, certainly, but Cohen's choice of subject matter ("that mundane object on which a philosopher might perch in order to think, ideation's unthought support; or in the palm, a spur to affect, cognition, and contemplation"[157]) seems inherently anthrodecentric in that it brings such an apparently bland object to the forefront of the reader's mind.

Harman identifies two different tendencies in philosophy that deflect attention from objects and consequently interfere with alien phenomenology. He calls these tendencies undermining and overmining. Harman claims that most kinds of ontology attempt to explain objects by either reducing them to their most fundamental units (undermining) or reducing them to their effects on other objects, particularly the human mind (overmining).[158] For Harman, objects are not reducible to either of these extremes. Undermining might involve reducing objects to their atomic structure (or even to smaller objects), or, for example, asserting that individual objects emerge from a boundless *apeiron* as in some pre-Socratic views. While undermining claims that objects are "too shallow to be real", overmining claims that objects are "too deep to be real".[159] An example of this is idealism, in which the existence of objects outside the mind of a human is uncertain or irrelevant. But Harman also includes social constructionism, in which reality is found in language, discourse or power. Relational ontologies such as Latour's are also examples of overmining, since they reduce objects to nothing but their relations with other things.[160] Both these extremes deny the autonomy of objects like tables, cats and leprechauns, which are the foundation of Harman's philosophy. Harman also identifies a tendency that he calls "duomining", which is the strategy of both undermining and overmining objects at the same time. Harman argues that modern science duomines objects by simultaneously reducing downwards to the most basic units of matter and claiming that all objects are knowable through mathematisation.[161] There are numerous examples of undermining and overmining in this book. Because humans have built machines from very small and basic materials, it is easy to believe that the key to explaining the sensation, experience and behaviour of technological artefacts lies in examining smaller and smaller components.[162] This is undermining the object, and it may be seen in the perusal of circuit diagrams or source code. The technological artefact may be overmined by reducing it to its relations or to its role in cultural discourse or power structures. We also see overmining in some ways of attributing a quality of mind to technological artefacts, which explain machine experience and behaviour in terms of consciousness. The artefacts are thus reduced to flows of thought which account for things like movement and speech. Both overmining and undermining avoid consideration of the alien nature of objects. Reduction in either direction is reduction away from the inscrutable and unknowable experience of nonhumans and towards something that humans can more easily understand. For example, it is easier to explain the actions of Voyager 2 by reducing it to its fundamental units than to speak of it as a unified and agentic being. It is also easier to understand Voyager 2 in terms of its relations with other objects, like planets, photons, electromagnetic radiation and components of its structure such as sensors. While this book presents thinkers that either undermine or overmine objects (or both), it is hoped that the analysis, in the main, walks a middle path between these two extremes. This is an important part of alien phenomenology. It is not acceptable merely to avoid anthropocentrism in the description of events; the language used must reflect this ongoing commitment. It must produce an uncomfortable feeling of alienation.

Harman notes that art is the domain in which undermined and overmined qualities converge.[163] This brings us back to the convergence of OOO and art. Bogost argues that the job of the philosopher should be to perform ontology or, as he calls it, "carpentry".[164] We should be "tracing the exhaust of [things'] effects on the surrounding world and speculating about the coupling between that black noise and the experiences internal to an object."[165] OOO is a laboratory, a place to build knowledge through the juxtaposition of conventionally ignored viewpoints. Bogost describes his own work with the game *Cow Clicker*, an exercise in both game design and activism, as an example of enacting his concerns through practical, carpentry-like philosophy.[166] For Bogost, written work is obfuscating and inaccessible, and he speaks sentimentally of his own engagement with the field: "so much of object-oriented ontology is, for me, a reclamation of a sense of wonder often lost in childhood".[167]

In Harman's early work, politics takes a back seat to metaphysics. However later work has drawn a link between politics and OOO.[168] And despite OOO's commitment to anti-correlationism, there are elements of the metaphysics that lend themselves to particular anthropocentric research programs. The application of OOO ideas to ecology is particularly significant, as in Morton's study of ecology. Nonhumans constitute both human bodies and the world in which we are irreversibly embedded.[169] OOO is a valuable site for transdisciplinary work, such as the intersection of ecological thought and philosophy.

OOO ideas have proliferated in different fields, but sometimes in reaction to it rather than by embracing it without reservation. The core of OOO has met with resistance from some feminists. A book of collected essays, *Object-Oriented Feminism*, edited by Katherine Behar, discusses the points of difference between OOO and object-oriented feminism (OOF). OOF is a reaction to OOO that shares some of OOO's convictions while rejecting others. After all, as Behar points out, groups such as women, people of colour, and the poor have historically been considered tools "there for the using".[170] Behar refers to the unpaid work of housewives: "Tools behave nicely. They are demure. They present themselves for service retiringly, to be used without reward in contrast to the work."[171] The woman who demands remuneration for her work is a "broken tool."[172] Objectification is not a novel experience for all humans. In my view, OOO breaks down the subject/object dualism, but does not go far enough in breaking down other dualisms, since breaking down dualisms is presumably not a high priority for most nonhumans. OOF also comments on the notably sexualised language used by writers like Harman, and plays with the erotic overtones, such as in this passage from Frenchy Lunning's essay:

> Despite Harman's assertion of ensnarement, these objects made of severed qualities never totally consummate their attraction. Instead, they remain forever in foreplay, in flux, in desire, and sometimes, in disgust. And it is in that strange alchemical exchange that the metaphor is successfully achieved.[173]

This essay juxtaposes Harman's concept of allure and Julia Kristeva's work on abjection. Even Harman's discussion of allure in *Guerrilla Metaphysics* implies a coquettish quality in metaphors, but OOF unapologetically emphasises what is left unsaid by Harman.[174]

OOO provides an ontology that invites comparison with a search engine. A search engine is sometimes erroneously thought to treat all inquiries the same. Any inquiry is valid, and results may be surprising and varied. However, in actuality, results appear based on "relevance", and the relevance may be determined by factors outside of the inquiry's control, such as censorship, sponsorship, popularity, and the user's previous search history. Alien phenomenology performs the same imperfect task. Objects are explicitly on an equal footing and we expect an exhaustive and impartial description. The carpentry of tools to explore the phenomenology of objects is based upon this impartiality. But humans who perform alien phenomenology are susceptible to biases. Just as Google's algorithm remains a secret, we cannot expect any one author to have certain knowledge of their own predispositions. This is why input from multiple authors is important. Objects withdraw, and we are left with only the sensual objects to which we are open. Just as a stone cannot perceive colour, an author with no training in medieval studies could not arrive at Cohen's conclusions.

Technology is not necessarily the focus of OOO, and there is no immediate link besides its origins in the tool analysis. OOO is metaphysics after all, and maintains the conviction that *all* objects are sites of wonder for those who choose to investigate them, not just marvellous objects. Yet there is a flavour to OOO that invites the discussion of technology. In many ways it is reminiscent of the idea of the cyborg. A cyborg comprises of technological parts and organic parts. Yet it is irreducible to either of these parts. The cyborg object becomes an object in its own right, able to form certain kinds of relations and totally unable to form others. The same is true of anthropomorphic machines. As humans, we are able to identify certain qualities in machines that are more like us or more alien. But there is no sense in which those anthropomorphic parts form relations in one way and alien parts form relations in another way. The human feelings they create, from emotional connection to a sense of the uncanny, belong to the whole object.

As OOO matures, it is increasingly mobilised politically, and therefore less inherently reliant on flatness. But we were always already unable to speak flatly. Jane Bennett has said that she believes the target of OOO to be "human hubris", the rejection of humans' search for truth and belief in reason.[175] Latterly, what she calls the "ethical impetus" of OOO has become more apparent.[176] Harman argues that OOO is increasingly relevant in a post-truth world, because it denies the possibility of true understanding of any object, despite insisting upon the object's reality.[177] OOO confronts the concept of truth in investigation as problematic from the beginning. So rather than insist upon any kind of flat analysis, or lament the need to speak in an anthropocentric way, we should make the socio-political implications of the analysis the focus of the inquiry. OOO analyses will always be political, and rather than ignore this we should emphasise it.

OOO also rejects human-centred politics, "which treats the political sphere as if it were the product of human nature and a purely human history".[178] Because of its implications for the flatness of ontology, nonhumans are allowed to take centre stage whenever OOO is used to discuss a social or political problem. This means that the study of technology using OOO is unusual. Technology has often been considered a special category of nonhuman objects, an idea that is in conflict with the principles of flat ontology, which rejects human-imposed hierarchies and groupings. However, flat ontology is not the final step in creating useful philosophical theory.[179] There must be some human creation of taxonomies of objects.

OOO is useful for the study of technology because objects are in the foreground, rather than existing in the background of scientific or social inquiry. In the interpretation of OOO presented in this book, technology is just as real and just as important as humanity – or at least a technological object is important for the technological object. Studying technology with OOO also permits an analysis of artefacts that includes the social and cultural background of that artefact, it just does so without privileging the social over other types of relations. A technological artefact may not require a different ontological analysis than a river or an electron. They are all equally objects. But technological artefacts are created by humans, and thus are strongly in relation with objects like ideas, platforms, and institutions, as well as things like metals, light, and individual humans.

Alien phenomenology focusses on individual objects, not vast groupings of objects. "Technology" is a term of convenience, but meaningless when considered from the perspective of each individual technological artefact. "Anthropomorphic machines", despite being a term used frequently in the book, is resisted as an ontologically significant category. Each of the case studies presented is anthropomorphic in its own way, both visually and functionally. In the case of SHRDLU, the specific incarnation of the object of study is available to us only through relations with a computer. The concept of the gynoid sexbot relates strongly to various different objects but is inherently incorporeal. They are both dramatically different from embodied objects like Grey Walter's robot tortoises. And as shown in the next chapter, the Microsoft Kinect's embodied status comes into conflict with its disembodied brand. Anthropomorphic machines may be grouped together for the sake of human convenience, but the grouping means little to the individual machines. Anthropomorphism is very important to us, but not to the machine itself. The machine really exists in its own right, and it may not experience humans or humanity as special objects with which it has a connection. OOO gives us this dialectical capacity: to rigorously attend to our own relationship with machines while holding in mind the alien nature of nonhumans. This is an unconventional approach to categories of technological artefacts, but is a part of the method for alienation that this book proposes.

This book apparently commits the very error Bogost warns against in privileging written work over carpentry. But perhaps it will inspire carpentry in future work embedded in the design, construction or exhibition of anthropomorphic machines.

Chapter 2
Kinect and "Natural" Interfaces

> What else does Kinect see?
>
> Kinect settings menu, Xbox One
> Accessed June 8 2017

The first case study of this book is the Microsoft Kinect, a natural user interface (NUI) with many remarkable aspects, not the least of which is its relative prevalence in living rooms throughout the developed world. There are two versions: Kinect 1.0 was released in 2010 for use with the Xbox 360 console and Kinect 2.0 was released concurrently with the Xbox One console in 2013.[180] Although initially bundled with all Xbox One consoles, Microsoft has since announced that the Kinect is no longer compulsory for all Xbox Ones in a move widely described as the "death" of the Kinect.[181] The story of the Kinect is one that highlights the inherent risk of deliberate anthropomorphism in technology for both users and corporations.

The performance of the Kinect presents a useful example for anyone interested in the strangeness of nonhuman worlds in a strongly human context. This chapter aims to arrive at an alien phenomenology of the Kinect that emphasises the unhuman agency and the contrived humanity of this apparently bland consumer product. Investigating agency means investigating the sensory world of the object, since an agent can only respond to things that it senses, which means that the sensory world of the

Kinect is very important for understanding the relationship that it has with humans. This chapter performs similar work to Adam Rothstein in *Drone* (mentioned in Chapter 1). The historical contexts available to this book (due to the author's disciplinary training) are incorporated with what Rothstein calls the "anatomy"[182] of the technological artefact. As emphasised in the previous chapter, this account can only ever be partial and anthropocentric. But as will be explained, collective alien phenomenology gives a more complete (if more confusing) view.

The complexity of the Kinect as it exists both in its individual and branded contexts demands a comprehensive explanation of how this artefact can sense, perceive, think, and act. This chapter argues that the Kinect *brand* is a distinct object which relates with objects like news articles, user forums, and advertisements; it is separate but in strong relation with any physical Kinect artefact. There are two different things to interest us here: a physically embodied sensor and a concept. Each is equally an object, and each is radically separate from the other. For OOO the absence of a physical form does not affect an object's ontological status. This chapter aims to argue for the existence of this division, then to investigate each object. There is an important complication to this analysis, which is that the brand and the individual unit have a very close relationship. That is perhaps why we humans tend to use the same word to refer to each object. "Kinect" can mean either the artefact or the brand. This chapter will attempt to discern one from the other, marking where the corporeal artefact ends and the incorporeal brand begins. This is something of an experiment, but it has a bearing on the case studies of the next chapters, all of which in some way depend upon the possibility of discerning different objects within the same linguistic umbrella.

The chaotic methodology of this chapter is representative of the transdisciplinary nature of alien phenomenology. Both artefact and brand are discussed separately and together. Because they are such different objects, this needs to be done through the lenses of different disciplines. The embodied artefact must be discussed in technical terms; what exactly is the Kinect, and what are the physical qualities that affect its relations with the world? Most of our information about this comes from Microsoft, which put some effort into explaining the technical capabilities of the

Kinect in 2012 and 2013, from the type of sensors that the NUI uses and the kinds of inputs it responds to. Since the approach of this chapter is derived from alien phenomenology, it will not skip over these vital details. Only by knowing more about the qualities of the object of study can we gain an appreciation of its most significant relations in its experiential world. The embodied Kinect is quite opaque for the average user, with its sensory capabilities quite difficult to comprehend. Learning about the way the Kinect relates to its world can tell us something about the way it experiences its world, and we acquire data on these relations by accessing and evaluating other people's observations. With the Kinect, we do this with user forums and trouble-shooting guides. This is a case of collective alien phenomenology, the struggles of ordinary human users to investigate the Kinect's sensory world through pseudoscientific collaboration.

The other object of study, the brand, is an incorporeal concept. Investigating the qualities of the brand requires different tools. The brand has strong relationships with very different objects from those that the artefact relates to. An artefact's sensory world comprises the immediate physical world, like the table on which it sits. The brand's sensory world tends to feature things like texts and other concepts.

There is a need for a degree of anthropomorphic language in this chapter, which will be exaggerated as much as possible in order to highlight the alien in the Kinect-artefact and Kinect-brand. Human phrases sometimes seem to readily apply to nonhumans and in setting up a multiplayer game one human might say "The Kinect can't see you", prompting the players to reorganise player positions in front of the television. As discussed in the introduction, such anthropomorphic phrasings used in daily life can be useful for increasing the tension between artefact and human reader, which aids in the juxtaposition between alien object and the language used to describe it. However, care must be taken in this analysis not to privilege the human in relation with the Kinect, since different interactions may be more salient for the Kinect at different times. Light bouncing off human bodies may not be the most important thing in the Kinect's sensory world at any given moment, although analyses of media often focus on this relationship. In addition, the cultural position of the Kinect is a significant factor in human-Kinect relations, but as far as possible these will not be prioritised over those factors that are significant for other objects that form relations with the Kinect. In fact, objects from the domain of cultural criticism come into close relation with objects from the domain of engineering in this analysis, and neither is privileged over the other.

We may not discover what it's like to be a Kinect, but we may gain insight into the way that humans conjecture that Kinects might sense, perceive and behave. Or perhaps more accurately, I will interpret the evidence of personal observation, technical accounts and reports by other users, and I will present this through the lens of my own view of the world. Unavoidably, you are reading a product of my relationship with the Kinect.

The objects themselves

NUIs

Natural user interfaces (NUIs) are a class of entities that are built with the aim of allowing humans to interact more "naturally" with machines. As a taxonomic group the class is rather problematic, since the "natural" aspects of the interaction are always in the eye of the human beholder. Two of the most important interfaces in NUIs encode gestures and natural speech, which are processes that do indeed come naturally to most humans but are contrived processes in machines. Gestures and natural speech must undergo a translation process in order to make sense to a machine. The NUI is an ideal to be strived for, but it is an anthropocentric ideal that erases the alien qualities of technological artefacts. NUIs in general are engineered to produce interfaces that are natural for humans, but the engineering challenges posed by capturing human gestures and speech result in interfaces that frequently fail to live up to the hype. Moreover, while more and more natural interaction with machines is attempted, natural speech and seamless gestural control remain elusive for the time being. Nevertheless, the idea of humans interacting "naturally" with machines is a popular one, and in 2014 Microsoft tied the future of its console line to the Kinect, which is designed to engage with both humans' gestural and speech-based "natural" modes of communication. To date, many entirely Kinect-based games exist but fewer have appeared over time. (There are also Kinect games that make use of traditional controllers in addition to speech or motion capture.) In support of this "natural" rhetoric, a team of animal-computer interaction (ACI, rather than HCI) researchers has introduced the Kinect to a group of orang-utans at Melbourne Zoo. Simple games on Kinect aim to enrich the animals' habitat and change the nature of encounters between humans and orang-utans.[183] Orang-utans do not play the same Kinect games that humans play in their homes, but the gestural interface is similar. This supports the view that all kinds of apes, including humans, are capable of learning to adapt to NUI technology quickly enough to make the technology accessible and usable. Another NUI designed for gaming interaction is Leap Motion, which is intended for the human use of virtual reality headsets.[184] A sensor for detecting hand motions is built into the headset so that users can interact with a virtual environment.

In addition to providing interfaces for entirely Kinect-based games, Microsoft and others have also been drawn to the possibilities allowed by "gestural excess" in controller-based games.[185] Gestural excess describes actions that do not contribute to success in a game but that are still performed by the player. Examples of gestural excess include recoiling from a blow in a fighting game and pumping one's fists in the air after a difficult win. The reactions are unnecessary because all the input to the machine comes through the controller and the console or computer's experience of the world does not extend to registering these enthusiastic human responses, but the responses are natural and often unconscious. Experiments in harnessing gestural excess have begun. For example, in the game *Kinect Sports Rivals* (KSR) wake racers can turn more quickly when they lean into a turn. Of course, once they become a part of the game the gestures are no longer excessive and become strategic ways to improve one's performance. This is an example of "grokking", what we might consider to be the opposite of gestural excess. Grokking a system means not merely understanding it but intuitively responding to it. The word "grok", first used by Robert A. Heinlein in his 1961 novel *Stranger in a Strange Land*, might describe the programmer's becoming very attuned and habituated to a programming language, or the relation that develops between a driver and a racing car. Humans learn the best way to communicate with nonhumans and, incidentally, gain knowledge of their qualities and way of relating to the world. In the Kinect's case grokking means a knowledge of gestural and spoken commands that the Kinect is able to sense, which then become second-nature to the user. It also means using the sensor in a way that works for the idiosyncratic home entertainment set-up, perhaps rather than the manner that is described in the manual. Movements come to be learned and rehearsed by humans until they lose the sense of being contrived or unnatural. Richard Harper and Helena Mentis identified this embodied comprehension in families, who did not know what "grokking" meant but still practised it, becoming adjusted to the system's shortcomings and quirks to the extent that they could take advantage of them.[186]

Other NUIs have been developed for purposes other than game console manipulation. A widely referenced (although fictitious) example of a NUI is the one used in the film *Minority Report* by Tom Cruise's character at the beginning of the film.[187] He organises images, zooms and makes mistakes with his gloved hands. The aesthetics of this moment have become iconic, both in interface design and in the cultural imaginary. The NUI in the film permits the quick manipulation of information by those who use skilful gestures. The NUI used in the film was in fact a real interface developed by John Underkoffler and controlled by gestures.[188] More recently, an interface using either Leap Motion or the Kinect has been trialled for use by radiologists and surgeons in sterile environments, the rationale being that staff can use the interface without risking infection in surgery patients.[189] The Kinect has also been used to develop user interfaces specifically for the use of elderly human users, who may not be digital natives and who may have difficulties with visual acuity and motor difficulties.[190] In the view of the researchers, "natural" interaction is particularly important for elderly people who may have difficulties with memory. Because it may be hard to remember specific hand gestures there must therefore be a reliance on the "instinct" of the human user.[191]

Don Norman's criticism of NUIs, entitled "Natural User Interfaces are Not Natural", points out that gestural systems are neither new nor groundbreaking.[192] He traces the history of digital non-contact gesture interfaces back to the invention of the theremin in 1928, which puts into question the impression of NUIs as futuristic technologies to be aspired to. He notes the similarity between modern NUIs and the first GUIs (graphical user interfaces). But as he explains, gesture-based NUIs are not merely as "natural" as GUIs are, they are actually even less intuitive because of a lack of feedback. Contemporary NUIs like the Kinect sometimes give feedback mostly by displaying images of the human bodies (and other objects) that they encounter,[193] while others provide feedback only by showing the effect of the gesture, such as by moving a videogame character's body in the same way as the human's.[194] In contrast, the use of a mouse provides immediate feedback to the human, who is following the path of the cursor on the screen with their eyes. But very few gesture-dependent interfaces deliver immediate feedback to the user that permits them to materially improve their use of the system. In addition, GUIs have become easier to use because a grammar of gestures has evolved over time. As an example, there is now a fairly universal text selection interaction. We can now see this grammar solidifying in haptic technologies as well. Norman's example is the scrolling action on smartphones. Over time, the "viscous friction" of the window has become standardised (in other words, the speed of the finger flick causes the scrolling to continue for a consistent length of time). Zooming has also become standardised. NUIs have not had the same penetration as touchscreens to date. The Kinect remains the only prevalent NUI on the market, both as an interface for the Xbox and for Windows PCs. But because the creation of Kinect games is no longer a priority for publishers (see below) there has been limited opportunity for Kinects to teach most of us how to communicate with them. Passing interaction with a Kinect permits it to teach humans basic rules, but thorough understanding of Kinects' experiences of the world and its way of feeding information back to the human user requires more extended sessions.

Natural language communication with machines is also an aspect of the ambition of NUIs, and also specifically of the Kinect. The Kinect is capable of listening for command words and prompting responses in the console, although as discussed below the communication between human and Kinect is not seamless or unbiased. A media archaeological study reveals an extended history of voice interaction in digital games, stretching back as far as the early 1980s in consoles and modules. Attempts to integrate voice interaction have come in waves, driven by new hardware and changes in how human users communicate and form relations with virtual characters.[195] Before the first attempts to commercialise voice interaction, there was a boom in attempting typed natural language communication with computers. SHRDLU, a natural language processing program built by Terry Winograd in the 1960s and 70s, is the subject of Chapter 4. The "natural" language used by the system pertained exclusively to the description and manipulation of a digital block world. But more recently the ambition of building NUIs that can understand and respond to humans has resulted in the creation and marketing of virtual assistants. Spoken communication with smartphones was popularised by Apple's Siri and is now replicated in other interfaces. Chatbots are seemingly ubiquitous. We are also witnessing the penetration of home assistants that exist purely to understand and respond to human speech. As shown later in this chapter, the Kinect has been accused of sexism, and the same is true of home assistants, although for different reasons. A home assistant that can not only listen but respond to questions must have a voice, and like human voices it must have an accent and the suggestion of a gender, which belies the presumed neutrality of home assistants.[196] Like motion capture, NLP technology has the implicit goal of creating nonhumans capable of interacting "naturally" with humans. It is still far from perfect in this regard. And the trouble with NLPs is that if they ever do become indistinguishable from human language users then they will be embroiled in the conflicts and struggles that surround and implicate all human voices.

Each part of the name "natural user interface" may be questioned from the perspective of the sensor itself. What is "natural" to the human is always engineered into the machine, so those aspects of the interface that are natural to the human are neutral for the artefact. A NUI constantly scans the environment for signs of human bodies and/or for sounds recognisable as human speech. In the case of the Kinect, "middleware" interprets these movements and sounds to decide what is relevant and what is not (see below). What is natural for a sensor is to detect everything that comes into its sensory field. The "user" part of the name is also problematic from the sensor's point of view. From the human perspective, we are the "users" of the NUI. But in a way, the sensor uses the human as well. Its program of action (to borrow an ANT term) is to detect images, sounds and depth, and to convey this information to the console. The sensor uses the human as a part of this environment. "User" in this book is typically qualified as "human user", but in accordance with OOO we must ask about the uses to which the sensor puts its environment. Finally, the NUI sensor is only part of the "interface", and relates to the console or computer, the screen, and the human. These are relevant issues to raise since the following account is of a NUI sensor and its way of being in the world.

Kinect-brand and Kinect-artefact

The Kinect has been mass-marketed relatively successfully. Originally given the codename "Project Natal", themes of revolution and innovation have been part of the Kinect mythology since its inception (it is also the name of a city in Brazil, following Microsoft's tendency to name projects after cities). Clearly the name "Kinect" is intended to evoke the "kinetic" motion-based play that represents a significant part of its interface, and "connect", which signifies connection and harmony with the game system and perhaps with other people.

The original Kinect was announced at E3 on June 1 2009 to a generally positive audience. Microsoft told the world there would be "no barriers and no learning curves"[197] when playing with Kinect 1.0, which was compatible only with the Xbox 360 console. It was released in 2010, several years after the Xbox 360 console which was launched at the end of 2005. The ambition of NUIs as a class was in evidence, such as in the slogan "You are the controller".[198] There was considerable speculation about what technology such as this would mean to gaming. The Kinect is used to augment videogame experiences, and some games are completely Kinect-based without the need for a traditional videogames controller like the standard Xbox One controller or non-traditional motion-based controllers such as the Playstation Move.

With the introduction of new hardware, Microsoft promoted a wider range of Kinect experiences for all Xbox users. The Xbox One was announced on 21st May 2013, as a successor to the Xbox 360 (in the media and gamer forums it was frequently referred to as "the Xbone", either affectionately or not so affectionately). It was released in November 2013. A Kinect 2.0 was bundled with every One sold, and they were featured prominently in pre-release advertisements, with claims that it would allow players to "reach into games and entertainment like never before."[199] The appearance of the Kinect 2.0 is of a long horizontal black bar supported on a stand. On one side is the Xbox logo, and on the other side is a circular cluster of sensors. Its minimalist design understates the sophistication of the sensor technology.

Other technological actants are important to define semantically in a study of Kinect. The term "console" will occasionally be used to denote the Xbox One console. The word generally refers to a computer specialised for videogames, typically suited for use with a television and with a games-oriented input device (or sometimes a handheld gaming device). Bernadette Flynn discusses the way that console gaming has eventually come to replace the family hearth, as the radio and television did before it.[200] Flynn's idea is particularly significant for studying the Kinect because there is often a necessary reorganisation of the living room to revolve around this "hearth" and the playing area in front of it. The Kinect is the main mode of input for some games, although often the menus can be navigated using the conventional controller and some games such as *Ryse: Son of Rome* use both input devices, sometimes simultaneously.[201] The console is the mediator for all of these objects: the playing space, the Kinect, the game disc, the installed content and the screen interface, facilitating their interaction to create the preconditions for play.

Further to this definition we must acknowledge the debate around the word "videogame". This is a topic that has been debated fiercely and with little resolution.[202] Definitions have changed and been contested for over sixty years, with definitions always seeming to leave out some electronic games and incorporating others.[203] Jesper Juul offers a definition that aims to bridge the gap between what seemed to be the inescapable debate between ludologists and narratologists.[204] According to him, games have six main qualities, including:

1. Rules
2. Variable, quantifiable outcome;
3. Value assigned to possible outcomes;
4. Player effort;
5. Player attached to outcome; and
6. Negotiable consequences.[205]

Bogost has called Juul's "an implicitly realist position about games, but a troubled one."²⁰⁶ This definition will suffice, however, as the relevant question here is what facilitates the videogame in its interaction with the Kinect. The videogame is not simply recorded on the physical disc or somehow within the digital download. These elements are necessary, but the videogame is also installed onto the console's hard drive. Therefore when we speak of a "videogame" we speak of a distributed entity (or actor-network) that links software, console operating system, sensors, the console's brand, the sensor's brand, hard drive, disc/downloaded entity, screen, input devices, the hands and brain of the player, internet presence as well as encompassing a concept that is influenced by other videogames and other kinds of games (such as board games) past and present.²⁰⁷ Lack of cooperation between human player and those cultural or material relations may cause disruption, such as when an input device lacks electric power or the manufacturer ceases to update the console. However, in a OOO sense the videogame is more than the sum of these components and their interactions. Viewing a videogame as simply a product of material and cultural influences constitute a duomining of the videogame, since parts of it are reduced to its material embodiment (undermining) and other parts are reduced to things like cultural impact and power relations (overmining). The videogame is undermined when we speak of it only as the impact of code on electronic signals. The videogame is overmined when we speak of it only as part of a broader cultural milieu. The videogame exists in relation with the hard drive, sensors and so on but these disparate elements also form part of the ontology of the videogame. The videogame is an integrated whole that acts in accordance with internal qualities, and relations are formed between the whole videogame and objects as diverse as consoles, humans, and online reviews.

Having defined the other closely related technological actants, we must turn to an articulation of what object is meant by the term "Kinect". For our purposes, the Kinect as a signifier relates both to the brand and to the artefact, which is an unsatisfactory arrangement for critical object studies. Yet neither brand nor artefact is more the Kinect than the other, and neither can be reduced to the other. The names of objects often hide the existence of several objects which, to humans, are simply described by a single label. To emphasise the importance of this distinction, for the purposes of this book the situated artefact will be labelled "Kinect-artefact" and the collective brand will be labelled "Kinect-brand". Initially, this distinction may seem reminiscent of Plato's theory of Forms.[208] We could conceive of the brand as a form that identifies certain qualities shared by all the Kinect-artefacts in the world: a "Kinect-ness". But the meaning of the term Kinect-brand presented here is quite different. The Kinect-brand is an object just as much as a Kinect-artefact (Harman has called brands "charismatic objects"[209]). It has its own qualities and relations. Often, as will be shown, these qualities and relations are significantly different from the qualities and relations of the Kinect-artefacts. The Kinect-brand is the ideal, untarnished imaginary that conveys certain ideas about the Kinect-artefact – such as those from Microsoft, and those from prominent users. Unlike the artefact, the Kinect-brand is abstracted and incorporeal. It carries with it the implicit understanding of its users and is a factor in Microsoft's planning process. Kinect-brand is the aspect of the machine that is present in popular culture, internet discussion and Microsoft presentations (unless the specificity of an individual Kinect is important). But it is not restricted to interactions with humans and the social since it acts and responds to other actants, such as the Microsoft corporation, retailers, regulatory authorities and so forth without forming one-to-one interactions with humans. As Scott Lash and Celia Lury put it, "the commodity is dead; the brand is alive."[210] By this they mean that the commodity (or artefact) is finished and final while the brand is self-modifying and possessing a memory. Although of course, as anyone who has ever owned a console knows, an artefact does not cease to be modified, either by software updates or by errant water spillages.

The Kinect-artefact is a material object: a plastic-encased set of sensors, including its middleware but ending, for lack of a better barrier, at the end of the cable connecting Kinect to console. The Kinect-artefact contains its knowledge of individual habits and interfaces with a specific Xbox One console. It is present in a way that the Kinect-brand is not; one buys a Kinect-brand but gets a Kinect-artefact. Neither object is comprehensively Kinect-like or human-like. It is important to note that this notation does not imply that the Kinect can be split into these two aspects: they are two *different* objects that both exist in the world, erroneously conflated into one by human (and specifically English) language conventions. It is also not the case that Kinect-artefacts are the real objects while Kinect-brands are sensual objects in OOO terms. The Kinect-brand is a real object too, with its own way of acting and its own way of sensing the world. The sensual object formed between Kinect-artefact and human is one in which light, sound, and movement play a key role. An individual human may also form a sensual object with the Kinect-brand, but it is likely to be through text and advertisements. Like all relations, it is unequal. The Kinect-brand may have a profound influence on the human, while most humans could only hope to make a tiny influence on the Kinect-brand, such as through the publication of an article, employment by Microsoft, or through activism.

It is difficult to arrive at a definition of either the Kinect-brand or the Kinect-artefact, but the signifier "Kinect" is also used by Microsoft to describe thousands of identical products. Each of these identical Kinect-artefacts shares similar Kinect-artefactish qualities, but they are in different relations. One Kinect-artefact might be in a strong relation with a table, another is still in its box. One Kinect-artefact can identify my face, another can identify yours. The term "Kinect" refers at once to one object and a distributed network of Kinect-brand and Kinect-artefact experiences. Treating the brand and artefact as two distinct objects is the only way to avoid blurring boundaries between the sensory relations of the more static artefact and the complex world of the idea of the object. And it also focusses our attention on the object itself, its inner essence, sensual connections with other objects and therefore how they touch and exert influence in their worlds. Of course, "artefact" and "brand" are very human-centric terms, but as the introduction to this book argues we may use human experiences to explain alien phenomena; in fact, this can strengthen our understanding.

The Kinect-brand

The Kinect-brand has evolved from an anthropocentric perception of a gimmicky (albeit entertaining) artefact to a more serious (albeit troubled) artefact. Kinect-brand 1.0 was originally marketed mostly to families. The portrayal of the technology facilitated by advertising and the assortment of available games signalled the intention of Microsoft to market the game to families and groups. This intended audience is hardly surprising given the history of gesture-based interfaces in gaming. The Nintendo Wii (2006) was strongly marketed to families, featuring games designed to appeal to all age groups as well as "games" like *Wii Fit*,[211] which used an electronic balance board to integrate fitness play into NUI use. The Eyetoy for Sony Playstation 2 (2003), and Playstation Eye for Sony Playstation 3 (2007), were aimed at children. One reason for the Kinect-brand's lasting association with children's games is that the physical nature of the play is well-suited to family and group interactions. Harper and Mentis suggest that the Kinect-artefact is used in households as a way of safely mocking others (particularly parents) when the system requires them to make some absurd movement or gesture.[212] In addition, the work of John Downs suggests that the role of the audience ("paraplay") becomes more varied when watching physical games, so there is greater engagement while waiting for a turn.[213] Another reason why the first Kinect was predominantly used in social and family gaming is because its detection of fine details was minimal. The responses to humans are more coarse-grained because the sensor either could not detect more detailed human movement, was unable to distinguish between humans and other objects, or because it could only communicate part of the information detected to the console. Social and family videogames are usually not so dependent on precision as are, for example, first-person shooters. The artefact and its way of being in the world exists in a bidirectional and asymmetrical relationship with the brand, determining the kind of human user to which the artefact is marketed. This, in turn, prompts the creation of artefacts that can be used effectively by children, families and groups. Coarse-grained relations between humans and the Kinect-artefact have had a gradual but inexorable effect on the Kinect-brand. Because the Kinect-artefact is so often (and so successfully) used by families or in other social settings, the Kinect-brand has become associated with this type of gameplay. But it has been argued that games with an affective component like *Kinectimals*,[214] which is aimed at young children, may be deployed for the didactic purpose of training human users,[215] which educates young people about the NUIs that are expected to become more prevalent in their lifetimes. The didactic induction of children into NUI use is a good way of shifting this cyclical relationship between artefacts and brands into eventual widespread adoption of NUIs like possible future

Kinect-artefacts. It is important to know this, because the Kinect-brand underwent a significant transformation when it came to relate to the Kinect-artefact 2.0.

The casual and family-friendly qualities of the Kinect-brand prior to Xbox One are evident from the kinds of games that were made for it. Games like *Kinectimals* make use of the technology by allowing users to reach into the virtual world and stroke pet animals. These and similar family games make use of Kinect-brand gimmicks, like the dubious promise of physically entering the game through body movements, but there are relatively few games that incorporate gestures into ordinary gameplay. There are many reasons for this, but the primary difficulty lay in the primitive nature of the technology (described below) which made it unsuitable for "hardcore" games. One online journalist opined:

> The Xbox 360's Kinect, Wii Remote and PlayStation Move are all rubbish and have spawned a handful of decent games between them, enveloped by piles of filthy, filthy shovelware.[216]

There are notable exceptions to this general rule. In 2011, Microsoft attempted to rebrand the Kinect-brand 1.0 for so-called "serious games".[217] This makes use of the rhetoric surrounding "casual" and "hardcore" games, and attempts to alter perceptions of the technology, presumably to raise the anticipation of the Xbox One and Kinect-brand 2.0. Some examples of hardcore games released specifically for the Kinect-brand 1.0 include survival horror game *Rise of Nightmares*,[218] which creates a creepy atmosphere by requiring the player to stand completely still, to swipe insects off virtual arms and to adopt a "fighting stance" to kill zombies; and *Fable: The Journey*,[219] which builds upon the popular *Fable* franchise and allows players to play sitting down, using gestures to attack monsters and other villains. By promoting these games, Microsoft aimed to associate the Kinect-brand with more mature themes and complex gameplay. In ANT terms, there was an attempt to enrol "hardcore" gamers into the Kinect-brand's network.[220] The artefact and its qualities remained the same but were repackaged for humans who were inclined to resist the idea that the Kinect-artefact could be useful or interesting for them. The "program" of the Kinect-brand resisted the "anti-program" of "hardcore" gamers by creating videogames that would counter the perception that Kinect-artefacts are only useful for casual games.[221] But the qualities of the Kinect-artefact made enrolment difficult. Both *Rise of Nightmares* and *Fable: The Journey* received mixed reviews.[222] These mediocre responses are typically blamed on a Kinect-artefact 1.0's slow reaction time and its difficulty in picking up subtle movements.[223] Developers also tended to incorrectly estimate the capabilities of the sensor. This is a good example of the misjudged phenomenology of an artefact causing it to form undesirable relations with cultural objects. Often, the games were made to show off the capabilities of the Kinect-artefact, rather than to make a good game that uses whatever hardware is available. This is likely due in part to the limited penetration of the Kinect-artefact into gamers' homes, which means that the market was small and there was little money to be spent on developing new games. This is a curious circumstance. The gamer's relationship with the earlier Kinect-artefact is one in which the machine was often privileged above the human's enjoyment for the purposes of building expectations for the forthcoming Kinect-artefact 2.0. Steve Woolgar describes this kind of thinking in "Configuring the user".[224] Users do not necessarily know what they want, because they are not attuned to the anticipated future of the technology, or because they desire a particular feature but are not prepared to pay for it.[225] Human users of the first Kinect-artefact did not possess the same information as Microsoft developers, such as the kinds of gaming experiences that would become possible with Kinect-artefact 2.0. However, the human-Kinect-artefact relation had an effect on the Kinect-brand, one that is contrary to that which was presumably intended. A poor understanding of the Kinect-artefact's phenomenology, either by the developers or the users (or both), meant that these "serious" games tended to damage the Kinect-brand from Microsoft and other developers' points of view. This perception of Kinect 1.0 set the scene for Kinect 2.0's release.

Another problem for the Kinect-brand is the widespread belief that the 2013 Kinect-artefacts could be used for covert surveillance, further privileging other interests above those of the gamer. When the Xbox One was first announced, it was revealed that Kinect-artefact 2.0s would need to be plugged in and active before the console could be used properly. This sparked considerable anger with gamers, and some bloggers and journalists worried in the aftermath of the announcement that Kinect-artefacts would represent a serious invasion of privacy, especially since the new sensors are powerful enough to detect heartbeats and can distinguish between individual voices.[226] This response was likely amplified by Edward Snowden's exposure of PRISM and its large-scale surveillance activities which came to light only a week after Microsoft's announcement, and also the subsequent allegation that Microsoft had provided the NSA with the ability to access email and other user applications.[227]

After these revelations became public, Microsoft entered a period of damage-control which permitted them to disassociate the console from the new concerns surrounding the sensors. The Kinect-brand needed to change its relationship with the Kinect-artefact. Microsoft assured gamers that Kinect-artefacts can be switched off when the console is not in use, but comparisons to Big Brother still abounded.[228] The influential gamer-themed webcomic *Penny Arcade* called the Kinect-artefact "the mandatory evil camera".[229] (Indeed, in February 2014 it was discovered that this fear was quite justified after the revelation that a UK intelligence agency had allegedly monitored webcam chat for several years and had considered using the new Kinect-artefact.[230]) This and other criticisms of the console meant that the announcement of the new system was surrounded by anger and confusion.[231] The Kinect-brand swiftly shifted stance from promoting the impressive sensory capabilities of the sensor to distancing itself from privacy concerns.[232]

Microsoft promised to remove other controversial features such as the mandatory DRM technology.[233] But despite the violent reaction from major videogames commentators, Microsoft did not announce that the Xbox One could be purchased without a Kinect-artefact. They explained that the high level of integration of Kinect-artefacts into the operations of the console meant that it was now impossible to sell it separately, and that they wanted the console to be consistent for developers. Maybe Microsoft did think the use of NUI technology could enhance traditional videogames, but to many it seemed that the self-described "hardcore" fanbase of the Xbox 360 would be obliged to pay more money for a device that was almost universally considered only suitable for casual and family games. Even if there was little miscommunication between the Kinect-artefact and the human, the

Kinect-brand was standing between them. Humans would almost never enter into a fresh and untarnished relation with a Kinect-artefact, free from concerns about privacy and the possible loss of "hardcore" status, and the Kinect-artefact could not enter into a relation with certain humans because many would now be compelled to switch it off or unplug it. This and the Xbox line's apparent transition from a powerful gaming engine to a home entertainment hub with a focus on television integration and wide range of entertainment media could be considered a way of distancing the console from "gamer culture" and its sometimes impenetrable language, tropes, and skilful gestures.

With a yet more drastic severing of Xbox One from the Kinect-brand, Microsoft seemed to cut its losses with its console. After the poor reviews of its flagship game *Kinect Sports Rivals*, and a little under a year after its initial announcement of the new console, Microsoft issued a statement stating that the console would no longer be compulsorily bundled with a Kinect-artefact. The post, entitled "Delivering More Choices For Fans",[234] has been widely called the death of the Kinect. The decision has been widely condemned by those who already own the console but appreciated by those who now have the opportunity to buy the console at a reduced price. In any case, the lack of incentive for new users to purchase Kinect-artefacts has led to fewer games integrating gestural and spoken commands, and the identity of the Kinect-brand is becoming less and less relevant, although many games have been released since this so-called "death". Would-be developers for the Kinect-artefact have to convince gamers that their product is so good that it is worth purchasing a peripheral as well. As for the Kinect-artefacts, their experiences were significantly affected by the announcement. Those users who did away with their Kinect-artefact changed the way that their Kinect-artefact senses and experiences its world, causing it to enter into relations with the back of cupboards and landfills, and preventing its sensory organs from activating those electronic pathways that were previously so central to its way of being in the world.

Kinect-artefact 1.0

Let's turn now to the sensory worlds of those Kinect-artefacts. Kinect 1.0's sensory organs function in a way that resembles human sensation, although they are not the same structurally. Kinect-artefact 1.0 combines the use of two types of sensors to achieve something that its developers, speaking on behalf of the artefact, claim is similar to human vision; it has a camera for the detection of the visible light spectrum; and a second system that emits and records infrared signals. The infrared sensor allows the Kinect-artefact to gauge depth, making users' gestures three dimensional. It emits infrared light from one point, which bounces off the player (and any other objects) before returning and being interpreted by the infrared sensor.[235] The Kinect-artefact also uses stereo microphones with complex software to reduce the problems of sound reverberation common to a home theatre situation, enabling the use of voice commands.[236] The only feature possessed by a Kinect-artefact 1.0 but not a Kinect-artefact 2.0 is a responsive tilt motor built into the stand which permits the Kinect-artefact 1.0 to respond to changes in player height or position. The IR emission and tilt stand adjustment are the only two interactions it can initiate besides its communication with the Xbox 360 console. The Kinect-artefact 1.0 was intended to be an affordable but high-quality intuitive interface for casual consumers, but it was difficult to use and required the user to exaggerate their movements.[237] There were serious challenges to creating meaningful communication between human and machine. Kinect-artefact 1.0 acted in accordance with the internal qualities determined by its structure, which often brought it into conflict with humans who wished it to act in different ways. The Kinect-artefact 1.0 had an unfortunate effect on the Kinect-brand by failing to perform as well as the Kinect-brand had enthusiastically advertised.

Despite these issues, several individuals and groups tried to gain an appreciation of Kinect-artefact 1.0's phenomenal world. Early exhibitions of the Kinect-artefact emphasised the capture of human movements through the "stickman" or skeletal visualisation, a crude depiction of the Kinect-artefact's idea of human bodies (which is the only relationship most creators and consumers were really interested in). It is a representation of human body parts as seen and interpreted by the Kinect-artefact. In Kinect-artefact 1.0 the skeletons were quite rudimentary, with straight lines connecting dots representing joints. The simplicity of the representations highlights the impression that the Kinect-artefact 1.0 used very sophisticated, labour-intensive technology to contrive an image recognition that most humans could do easily. The Kinect-artefact, despite its efforts to be natural (read: human) struggled even to perform what would be a very simple task for most humans. Nevertheless, this visual was part of the campaign for Kinect-artefact 1.0 for Windows, part of the effort to encourage people to do their own things with their Kinect-artefact.

Stick-man is an attempt to code certain aspects of the Kinect-artefact's phenomenal world for human consumption. It gives the user or the engineer an insight into the machinic experience. However, it is not intended as an interactive art project. Rather it is a tool for those individuals who want to persuade the Kinect-artefact to do something. It is an expedient hijacking of Kinect-artefact sensory systems to ensure future compliance with human motion. Many impressive projects were attempted with the implicit or explicit blessing of Microsoft. Among them is Chris Vik's 2011 performance at Melbourne Town Hall. Vik used a special Kinect-artefact-based interface to control a 4-storey organ with hand gestures.[238]

However, there were serious problems with basic gameplay in the Kinect-artefact 1.0. As argued later in the chapter, the Kinect-artefact *wants* to communicate well with the user. But it can only ever react to events in accordance with its qualities. One significant problem was the Kinect-artefact failing to recognise humans when the lighting was wrong. A quality of the Kinect-artefact 1.0 vision system is that it recognises the presence of humans by the colour of the light that it receives through its lens, and because lighting affects the colour of light, the Kinect's interpretations were unpredictable to humans. The body mapping software was also too simplistic to capture all the subtleties of human movement and the gestures did not work as consistently as one might hope. Voice commands were not so ineffective but were underused by the software. These kinds of problems that the Kinect-artefact 1.0 struggled with are to some extent understandable to us, since humans also struggle to detect objects when conditions are too light or dark. Humans make errors in perception. But it is a little more difficult to say that we could mistake a person for a nonhuman object simply because they use a wheelchair, are moving too fast or are in poorly-lit conditions. We usually have little difficulty tracking the movement of a person's hand in space or knowing that a hand concealed behind a person's back is still there. These frustrating and sometimes inexplicable problems come from the extreme restriction of the Kinect-artefact 1.0's world. It has limitations to its sensory capabilities that mean it has real difficulty carrying out its program of action.

Despite all its limitations, the Kinect-artefact 1.0 remained popular through the three years prior to the end of the Xbox 360 console cycle, selling more than 24 million units.[239] We can readily identify what was beneficial about this arrangement for Microsoft: it made money by selling Kinect-brand games (even if it possibly lost money on the Kinect-artefact itself) and it set the stage for future success in the console wars. It sustained Microsoft through a longer-than-average console cycle brought on by the Global Financial Crisis. And it offered a tantalising look at what might be possible with the next Kinect-artefact.

Kinect-artefact 2.0

The advertising, statements from Microsoft, and sensationalist news stories[240] prior to the release of Kinect 2.0 initially raised human user expectations. With the release of the Xbox One, Microsoft promised that the Kinect-artefact had been "completely re-engineered"[241], and seemed intent on reversing the popular perception of the device as a gimmick. The sensor now had a much larger field of view, compensating for one of the Kinect-artefact 1.0's biggest problems. The lighting problems were also fixed by adding imaging technology that does not depend on visible light. Both cameras working in concert create a more complete image of the user. It seemed that Kinect-artefact 2.0 would be better able to achieve its goals than Kinect-artefact 1.0, to the overall benefit of the Kinect-brand.

The Kinect-artefact 2.0, unlike the Xbox 360 model, uses a time-of-flight (TOF) camera. It measures the round trip time (RTT) of photons emitted in flashes from the Kinect-artefact.[242] The photons bounce off objects and return to the Kinect-artefact, from which the distance of objects can be measured using the speed of light constant. This creates a 3D image of whatever objects are positioned in front of the Kinect-artefact, and in far greater detail than that offered by a Kinect-artefact 1.0. Added to the 1080p visible light camera there is also an active IR sensor that addresses the lighting problems experienced with the Kinect-artefact 1.0. With a

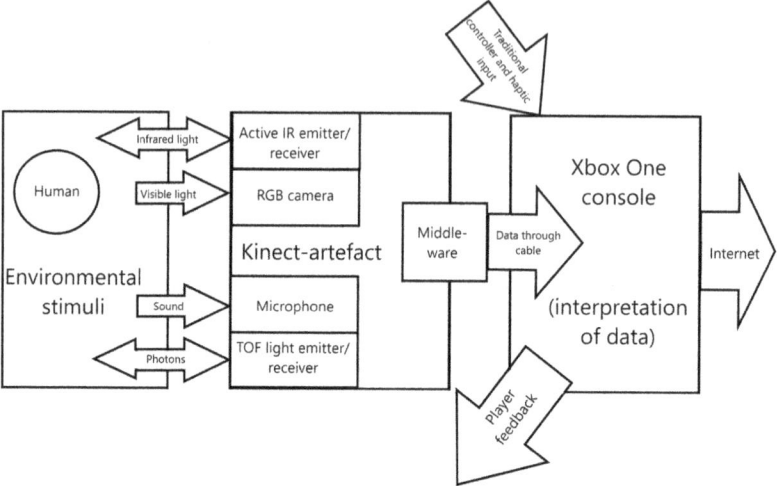

Figure 1: Diagram of a Kinect-artefact 2.0's basic technical relationship to other objects.

Kinect-artefact 1.0, too much lighting on one side of the room could adversely affect the sensor's ability to pick up players. But the new sensor not only fixed this problem, it allowed users to interact with the sensor in total darkness.

Two pre-launch videos made by Microsoft were entitled "Behind the eyes of Xbox One Kinect" and "Inside the brains of Xbox One Kinect",[243] names that continue Microsoft's efforts to anthropomorphise the Kinect-brand and to encourage gamers and users to think of their product as an active, smart participant in creating entertainment experiences.[244] The videos provide visual representations of one part of the Kinect-artefact 2.0's world: its experience of the human and the human's engagement with the console and software. They show how the Kinect-artefact integrates TOF cameras, active IR and visible light cameras in different ways. Most of the focus of the videos is on how the Kinect-artefact can sense and perceive human bodies and eliminate other objects. A team of developers demonstrated this using their own bodies. The videos all came from a sample of a session created and published by Microsoft Research, in which humans moved around and demonstrated the capabilities of the Kinect-artefact in use. The room in which the video was recorded is quite large with few obstacles. This permitted the participants in the video a wide range of movement. Presumably the participants were already very familiar with the Kinect-artefact's abilities and deficiencies, and they were able to use the Kinect-artefact skilfully, possibly with a history of grokking.

Firstly, there was an image of the human body as the Kinect-artefact "sees" in total darkness Next, there was a coloured skeleton image, which was evidently more complex than the stick-man images of the Kinect-artefact 1.0. The Kinect-artefact 2.0's skeletal tracking has greater articulation of the hips, spine and shoulders as well as recognition of thumbs and the length of the hand. In addition to the skeleton, what the developer teams called "block-man" revealed how the Kinect-artefact sees rotation in the joints and how body parts relate to each other. Instead of straight lines, major parts of the human body (head, shoulder, lower leg, hand, etc.) were depicted as coloured blocks. Block-man showed the orientation or "rotation" of the body parts.

Developer Kareem Choudry claims in "Inside the brains of Xbox One Kinect" that the skeleton is only a small part of the story when it comes to mapping the body. "If you really want to understand what's going on with the human you need to understand and describe what's going on with the muscles, the forces, the torque." The next image in the video is "muscle-man", and it depicts how the Kinect-artefact 2.0 thinks the muscles in the human are behaving. Muscle-man is similar to the skeleton image, but with more rounded muscle-like lines representing limbs, torso, and head. The body parts are coloured on a spectrum from red to green; green represents "no force" and red represents high intensity. A leg carrying more of the human's weight is darker in colour, and circles under the feet represent impact on the floor. When the individual is punching the air, as one might do in a fighting game, the power of their blow is represented with a circle of varying size. While the skeleton and the block-man are merely closer to the ways that humans perceive bodies, muscle-man is a mode of perception that the Kinect-artefact has which humans lack. Similarly, the Kinect-artefact uses its alien senses to detect exertion in humans. The visible light spectrum camera and the Active IR camera work together to deduce the user's heart rate by detecting tiny changes on the skin. This is a type of sensation that humans do not experience, although they are typically able to draw similar conclusions to the Kinect-artefact (i.e. "this human has recently exerted themselves").

These visual models are a good representation of what the Xbox deduces about human bodies from its "eyes": the colour camera, active IR sensor and time-of-flight sensor. The videos are, however, of varying utility to exploring the phenomenal world of the Kinect-artefact. They are attempts to render human figures in human terms by first translating them through an alien interface. The videos are, after all, promotional images and reflect the "brain" of the Kinect-artefact. The "brain" is located in the middleware of the Kinect-artefact that makes sense of what the sensors detect – identifying certain patterns as human bodies, for instance.

The Kinect-artefact is an extremely human-oriented device, both in its sensation by the sensors and the mediation by the middleware. Its sensors were designed to emit and receive radiation that is best able to give information about human bodies, as well as certain other things such as the floor the human stands on and the controller the human holds. The middleware takes these already-biased observations and further exaggerates the body positions before relaying the information to the console. Not all the information picked up by the sensors is used. From the representations shown in the promotional videos it seems clear that the Kinect-artefact is not interested in the position of furniture or what is hung on the walls, beyond its role in obstructing the Kinect-artefact's vision. The Kinect-artefact is oriented to the human. And yet, it would be difficult to claim that the Kinect-artefact is human-like in its sensing and perceiving. It is an extraordinarily alien object, even for a videogames accessory. It constructs a view of the human using bursts of photons and block-like models that are quite foreign to human vision; it receives input from one direction but can only send data onwards to the console; and it can be put into a state in which its sensors do not respond at all.

The phenomenology of nonhumans is ever more pertinent to everyday life. The Internet of Things project aims to construct a new internet-based way of living as that of a planet so enriched with interlinked sensors that it is capable of far more advanced coordination than previously. The Kinect-artefact is a sensor that affords a different kind of relation between human and console. But can we really only describe a Kinect-artefact-console hybrid as a sensor? In a personification of the phenomenon we would surely describe it as a sensory-perceptive entity. But the concept of perception, when applied to a nonhuman, is a problematic one for the information processing theory of knowledge.[245] The interiority of machines is questionable, and their irreducibility seems absurd when each component is created by humans. This will be addressed further in the next chapter, with the discussion of Umwelten and consciousness, but briefly, it seems evident from the way by which the Kinect-artefact conveys and transforms information from human to console that it *at least possesses some interior qualities*. These are the qualities that cause it to relate in certain ways – Kinect-artefactish ways. Consequently, if we put aside for a moment questions of mind and consciousness, the study of the alien phenomenology of the Kinect-artefact is valid. And due to its status as a contentious cultural object, its phenomenology is of particular interest to humans.

Communication, sensation and relation of the Kinect-artefact

At the time of release, Microsoft billed Kinect-artefact technology as something that makes interaction with the Xbox One more natural. The language of "eyes" and "brains" conveys the impression that the Kinect-artefact has an implicit humanity so that humans may communicate with it more easily. The human need not learn and adapt to use the system, because the system has been adapted to be human-like. The Kinect-artefact participates in semiotic interaction with multiple objects, most notably the console, but also with humans, furniture, the speed of light constant and the Kinect-brand. Human relations with a Kinect-artefact are affected by all of its other relations with the world, and it is not always clear why. Under these conditions, humans often band together to investigate an object in an informal and practical way, as in the case of the "Waking up" problem.

"Waking up" the Kinect-artefact

In the immediate aftermath of the release of the Xbox One, corporation and customers came to a gradual equilibrium with regards to complaints and suggestions about Kinect-artefact 2.0. Several major bugs and design problems were reported to Microsoft and resolved. Difficulties for Kinect-artefacts included identifying humans and conveying information about humans (and their surroundings) to the console, and the console had difficulty interpreting that information. Software updates made it easier for Kinect-artefacts to behave in a way that was more pleasing to their humans, improving the relationship. However, many problems remained. One example is that the furniture in the room containing the Kinect-artefact must be carefully arranged. The Kinect-artefact is widely described as "finicky" or discriminatory, and the lack of a general understanding of its phenomenology has resulted in an abundance of pseudoscientific explanations for its behaviour. Humans needed to understand what the Kinect-artefact wanted from its environment before they could form meaningful semiotic interactions with it, and eventually grok the system. The most prominent issue after the Kinect-artefact 2.0's release is known as the "Xbox On" problem, and it represents a starting point for engaging with the phenomenal world of the Kinect-artefact. The data for this section was gathered from the official online Xbox One support forum[246] and from Reddit's Kinect and Xbox One subreddits.[247]

The "Xbox On" problem has become an emblematic instance of the communication problems between human and Kinect-artefact. A major emphasis for this generation of the Kinect-brand is on voice commands, largely replacing the hated gesture controls of the Xbox 360. Kinect-artefacts (those that are able to hear and understand human voices) convey that information to the console and thus are able to control many aspects of the interface. But Kinect-artefacts often have trouble understanding human intent through language, and this causes humans to come into conflict with the Kinect-artefact as well as the Kinect-brand. To use Heidegger's terms, when ready-to-hand for the human the Kinect-artefact hears, understands, and acts. It can become present-at-hand for the human by failing to hear, understand, or act. A user can play movies without needing to touch a controller, and they may perform all the additional playback commands with voice as well. Kinect-brand games and many major menu operations can be activated with speech. These features have been welcomed by users with physical disabilities (specifically blind people or people with low mobility), pregnant women and parents with small children, and are reported to be useful when cooking and while sick, since without voice controls the Xbox One cannot respond to immediate human needs without input from a controller. One user reported being able to use a voice command to turn the Xbox One on from the ground floor so that it would power up while they were making their way down to the basement. The Xbox One also learns the cadences of individual users' speech so that over time it responds more easily to voice commands.

But there is one problem which has been extensively identified: the difficulty encountered by many users trying to switch their console on with the "Xbox On" command. When the console is on "instant-on" standby, it can reactivate in response to a human uttering the voice command. In the Microsoft advertising materials for the new Kinect-brand, this feature was given particular prominence. On its website, it claims that "Xbox One was designed for today's fast-paced lifestyle. It wakes up instantly when you say, 'Xbox On,' and even turns on your TV."[248] But unfortunately many users find it difficult or impossible to wake the console up without physically touching the box or controller, leaving them wondering whether the Kinect-artefact is poorly designed or whether their environment is simply not convivial for a Kinect-artefact.[249] This is an important instance in which the Kinect-brand becomes disassociated with the Kinect-artefact, which simply cannot perform what the Kinect-brand has promised. The Kinect-artefact does not hear, understand, or act on the voice command. Identifying why this happens requires learning about the Kinect's way of being in the world. It becomes present-at-hand for the human, and it becomes the subject of alien phenomenology.

The desire to switch the console on without needing to touch a controller is a practical concern. But those users that found difficulties with this command exhibited a frustration and disappointment that may be judged excessive for a practical issue. Several commenters expressed a sense of "betrayal" after buying consoles with the belief that it would work like it does in its advertisements and subsequently spending considerable time and effort trying to make the feature work. There was a particular anger because users felt they were "forced" to buy the Kinect-artefact with the console despite the widespread privacy concerns. There was also a feeling of being ignored by Microsoft and particularly of support staff failing to read the forums and therefore not realising that there was a problem, or perhaps even that Microsoft was refusing to admit that there was a problem. This may have been a residual sentiment from Microsoft's failure to acknowledge the "Red Ring of Death" problem, a malfunction that rendered Xbox 360s unusable.[250] Presumably, the Kinect-brand also experienced a sort of frustration from this issue. It threatened the relationship between Kinect-brand and existing users of the Xbox One console, which may have made it less likely that other users would purchase a Kinect-artefact. User forums and social media comments may be in a strong relationship with the Kinect-brand, and at the time of this problem these online influencers of opinion were consumed with the "Waking up the Kinect" issue. The sentiments of the posters leave the impression that there was something greater than a practical concern inherent to this difficulty, especially since other practical problems, such as the lack of a choice when using the Kinect-artefact for voice communication in multiplayer games, did not attract the same demands for a solution. It may be that this disproportionate response emerged from the portrayal of advanced computers in science fiction as well as in corporations' depiction of present-day technology as becoming closer to anticipated Artificial General Intelligence computers. There is a craving for anthropomorphism in machines. Patterns of sleep and wakefulness are among the most culturally deterministic aspects of our biological existence, and any kind of "smart" machine that approaches human intelligence should be wakened in a human way.

The community of users trying to find a solution to this problem explored many possible avenues. One interesting class of explanation suggests that users with certain types of voices had greater success than others. The main observation made by many users was that female voices were often able to activate the console while male voices could not (although some users reported the opposite effect). Some users reported success after uttering the command in a falsetto register. Many users had suggestions for the best pronunciation of words. A particularly common piece of advice in any region is to pronounce the word "on" like "ahn", with more of a North American accent. One user claimed that using a "really nasally 'nerd voice'" improved the voice commands. Some users claimed that it is better to say it very quickly, and others claimed that pausing between the two words might be an improvement. There was an awareness that Kinect-artefacts possess similar internal qualities that affect relationships – in other words, the beginnings of an informal assessment of the machine's experience of the world. Several users reported using a Youtube video to turn on the console, made by individuals whose voices were easily understood by Kinect-artefacts. At the time of my research there were two videos that were being used, both including the apparently male filmmakers turning on the console with their voices.[251] By sampling the recording, users were able to turn on their consoles using the voice of the filmmaker. One caused a particular sensation, since the filmmaker with a British accent was able to turn on previously recalcitrant North American systems. One possibility proposed is that a recording on an electronic device clips out the lower tones in a person's speech, supporting the theory that higher tones improve performance. The degree of indignation on this point reflects the predominance of male gamers as users of the previous Xbox consoles. The anger among early adopters came from feeling as though Microsoft owed "us" the best possible user interface and had not delivered.

There were also apparently problems with accents. The "Xbox On" command was not provided in Australia at launch; there was no option to put it in an instant-on energy level when it was switched off, so it needed to be switched on manually. It could be very simply accessed, however, by switching the console's region settings to one of the countries for which the "Xbox On" was available, such as USA, UK or Canada. This meant, however, that the Kinect-artefact was not able to comprehend an Australian accent all the time. Changing the region changes the language that a Kinect-artefact understands and responds to. An utterance made with a different accent does not always produce the desired effect. The human speaks Australian English, but the Kinect-artefact hears imperfect US English. These restrictions may make the "smart" object seem remarkably slow-witted, or possibly even sexist or racist.[252] Users are beginning to expect natural language processing that can communicate additional information, respond to jokes and understand idioms (Apple's Siri and Amazon's Alexa come to mind), which are all much more tied to local culture than to broad language groups.

The Kinect-artefact itself does not tell us explicitly what gender or nationality it prefers; users have been required to work it out by observing the positive and negative feedback given by the Kinect-artefact. Some users performed experiments to find ways of making contact. They engaged in an informal program of alien phenomenology, deducing the nature of Kinect-artefact sensation from the positive and negative feedback of the Kinect-artefacts. It is not clear whether the problem was in the position of the Kinect-artefact, the room acoustics, the calibration of the Kinect-artefact, the energy settings, or a hardware problem, but it is clear that the interaction of the Kinect-artefact with other nonhumans was a significant factor in whether it could communicate effectively with humans. Microsoft's advice for the Kinect-artefact microphone is fairly simple: follow the calibration procedure properly and make sure there is not too much noise in the room.[253] Yet on the forums there were a vast number of suggestions for ensuring that the Kinect-artefact works properly. Some users reported having completely moved their console and the rest of their home media ecology to another room, particularly smaller rooms or rooms with carpeting. A collective effort to find the ideal positioning of the Kinect-artefact was underway by humans, and Kinect-artefacts participated in this process by punishing and rewarding consistently, according to each artefact's internal qualities. The investigation into positioning included ideal distances between the Kinect-artefact and the human. One user recommended crouching down 4-5 feet away, another suggested two inches. The forums were full of appeals to describe users' home theatre set-ups, sensor positions and room design factors. Its ability to hear also appeared to change over time, depending on how long it had been switched off. The "sleep" metaphor seems apt, since at first its sleep appeared to be light but it then became less easily roused. The Kinect-artefact became sealed off from the humans attempting to make contact with it. In the absence of instruction from Microsoft or successful hacks of the Kinect-artefacts, the participants in this process, in short, adopted a sort of naïve alien phenomenology. The quiddity or interiority of the Kinect-artefact was assumed, and the users experimented with different kinds of objects to try to gain access to as much information as possible. But they could never *be* their Kinect-artefact and seemed to be repeatedly frustrated by failures to access the real root of the problem.

A common theme in the forums was describing the Kinect-artefact and Xbox One as though it were an animal to be tamed. One user suggested that the sensitive Xbox One needed a "caring, motherly voice", and many expressed the sentiment that the Kinect-artefact needed to be disciplined and educated. The Xbox One needed taming (or domestication[254]) because it entered into entanglements in pre-existing and highly heterogeneous living room set-ups. It must be nestled into the pre-established network of television, other consoles, cable television or set top box and peripheral gaming objects as well as becoming a focal point of all these activities. The Xbox One was intended to make these other electronics controllable through a single system (at least in the United States).

The Xbox One also domesticated its human user. Using negative and positive feedback, the human was encouraged to position the Kinect-artefact prominently in the room, to rearrange their furniture and alter their electronics set-up. They were even compelled to change the way they speak and move. The Kinect-artefact cannot give clear instructions to the console without this domestication and must encourage the human user to act in a way that engages with certain parts of the Kinect-artefact's way of being in the world.

Another part of this domestication involved encouraging the human to use the applications compatible with the Xbox One. In their book *Games of Empire*, Nick Dyer-Witherford and Greig de Peuter chart Microsoft's moves and describe them as processes of "deterritorialization and reterritorialization".[255] They were speaking of the two older Xbox consoles in addition to certain occurrences in the world of personal computing, but it may also be found in Microsoft's strategy with the Xbox One. The key is to create a multimedia platform that forms an entertainment hub in people's homes, from which additional services can be marketed and sold. The Microsoft-owned services included:

- Skype
- Xbox Video and Music (paid subscription or single payment libraries)
- Integrated Bing internet search
- Internet Explorer

- Xbox Fitness (purchasable work-out videos designed for Kinect-artefact integration)
- SkyDrive
- Upload Studio (allows users to create, modify and publish videos of gameplay).

As well as products that are not owned by Microsoft:

- Netflix
- Amazon Lovefilm
- Twitch (leading machinima and gameplay video platform) and Machinima (gamer and programming network)
- Youtube
- TED
- Hulu Plus in addition to many regional premium cable channels and video streaming apps.

The console comes to control the gaming, social, video and music services. Once the cable television, speakers and other elements of the home theatre are under the control of the console (or the Kinect-artefact) it becomes a vital component and cannot be abandoned. The user is encouraged to use Microsoft services on other platforms and perhaps even to buy Microsoft smartphones, tablets and to run Windows on their personal computers. The hardware and applications used for socialising, home entertainment, mobile entertainment, internet surfing, gaming and media library management are, in this metaphor, "territories" that Microsoft is endeavouring to capture in the everyday media consumption of its users. The domestication of the user by the Kinect-artefact allows Microsoft to capitalise on the prominent position of the Xbox One in homes. If one platform can provide so many different functions, and interacts so seamlessly with other software sold by the same company, why use products made by Google, Sony, or Apple? In the terminology of ANT, the platform becomes an obligatory passage point in the home entertainment system.[256] The human-object and object-object relations prevent any direct human-Microsoft relations, and the brand remains safely abstracted and uninterrogated. The Kinect-brand is like the lancet liver fluke *Dicrocoelium dendriticum*, which manipulates the nerves of ants (through unknown means) to compel them to leave their colony. As Jan Pechenik puts it, the ants are then "obliged to crawl upward on a blade of grass and to bite down firmly upon its tip".[257] They are unable to open their jaws during peak ruminant grazing times, and cattle, sheep, and pigs are thus more likely to consume the ant, and the fluke completes the next phase of its life cycle in the ruminant's digestive system. Many parasitic organisms change the behaviour of their hosts in similar ways. In this analogy, the Kinect-brand alters the brain of the human in subtle ways, compelling the human to change its behaviour in a way that better suits the brand. Of course, particular individuals in a species may develop resistances to parasites, or the parasite's genetic profile may drift, altering this optimal arrangement. Similarly, some humans are already inhabited by other parasites (such as those of the Google or Sony genus), develop a disinterest in the brand over time, or are so over-exposed to attempts at corporate manipulation that they do not cooperate with the brand.[258]

This is an important way of ensuring that users of the previous system adopt the next system. To some extent this tactic enabled Microsoft to retain a core of Xbox 360 users who did not wish to abandon their Xbox Achievements (rewards that may be earned by achieving certain goals during gameplay), avatar and profile, plus the rewards of Xbox Live Gold membership. There were also advantages for Playstation 3 users to upgrade to the Playstation 4, such as the retention of Trophies in the new system. But the Playstation 4 with its many other arguable advantages, lacked the totalising home theatre presence and was therefore less likely to become a permanent fixture in any home theatre set-up, especially given the increasingly long console cycles. Perhaps Microsoft believed that the generation of gamers that were loyal to the Xbox 360 for eight years had matured and wanted their gaming history to be preserved with the additional advantages of partner- and child-friendly features. The "Xbox On" command was a fundamental feature of this process of territorialisation. The user could perform the science fiction-like act of communicating with their computer through speech – a novelty which the core users represented on the forums appeared to enjoy. Furthermore, in regions where the process was supported, the "Xbox On" command awakened all the home theatre elements. The word "Xbox" here is meant to substitute "complete entertainment experience". The Xbox *is* domestic entertainment.

In addition to the logic of domestication in console expectations, the nature of machine learning and preponderance of "smart" devices has led to a set of impossible expectations for this generation of consoles. Microsoft is not the only company responsible for this impression, but it certainly cultivated the anticipation of a device that would have an agency and anthropomorphic intelligence well beyond previous consoles. Anthropomorphising machines is a powerful method for conveying the utility and sophistication of an electronic entity but is often highly misleading. In the case of the Xbox One, this tactic backfired and caused the impressions of treachery and deception that resounded in the forums. The trick of anthropomorphism was exposed.

Kinect Sports Rivals

Another re-negotiation of the line between humans and nonhumans is to be found in the game *Kinect Sports Rivals*[259], which turns humans into avatars. *Kinect Sports Rivals* (*KSR*) was one of the first titles to explicitly focus on the Kinect-brand. It is a sports-themed game in which gamers play several minigames: wake racing, bowling, soccer, sharp shooting, rock climbing and tennis. Originally planned for concurrent release with the Xbox One it was delayed until release on April 8 2014. The title was an opportunity for Microsoft to exhibit some of the new Kinect-artefact's power and potential, but the game also demonstrates a fascinating use of AI in collocated social gameplay. Gamers can play in multiplayer groups as in earlier Kinect-brand party games; up to four individuals can use a single system at once (and all may be logged into their personal Microsoft accounts). But an additional social element occurs online, either playing with an absent friend in real time or via the use of an AI avatar who can play against its user's friends without the gamer initiating or authorising the game. In other words, Player 1 can play against Player 2 even when Player 2 is asleep. Human and nonhuman can play in the same way as human and human, but with the added complication that the nonhuman mimics a specific human rather than a general one. *Forza 5* for Xbox One has a similar arrangement in which the human player can be represented by their "drivatar" AI simulation.[260]

The game is comparable to *Wii Sports*, a game released to showcase the Nintendo Wii in 2006.[261] *Wii Sports* is a similar game albeit much simpler both in ambiance and in gameplay. But because the pared-down Wiimote tended to respond to player intentions under the best conditions, the game was generally acknowledged to be a success. *KSR* was not as successful. Although the creative island setting is compelling, and the social functions work well, the game received mediocre reviews with a resounding tone of dissatisfaction at a sensor that had failed to live up to expectations for the second time in three years. The production company, Rare, has since been forced to lay off staff in the wake of the game's failure.[262] Perhaps this is because the standards for the game were set too high, as after the release of the demo *Kinect Sports Rivals Preseason* in November 2013,[263] many reviewers were optimistic about the game's prospects. Chris Carter of *Destructoid* reviewed *Preseason* and pronounced it "Surprisingly Not Terrible",[264] which is a significant endorsement from a "hardcore" games publication. The demo contained one single wake racing track with relatively intuitive controls and clearly showcased the best that the full game would offer. Most reviews acknowledge the wake racing as the most successful feature, while going on to say that some or all of the other minigames fail to replicate its elegant simplicity.

For the purposes of this book, the avatar creation and autonomy of the avatars used in *KSR* are the most interesting qualities. A detailed modelling of a human player into an electronic avatar aims to recreate the human player in the game.[265] This modelling is three-fold. Firstly, the creation of the avatar's physical form is performed, in surprising detail, by the Kinect-artefact's view of the player, particularly of the player's face (the game appears to be fairly diplomatic when estimating the player's body proportions). Microsoft's promotional materials claim that the Kinect-artefact "instantly scans you and captures your likeness as a champion".[266] The avatar is, of course, something of a caricature, the game having selected those most distinctive or significant features in a human's face. Secondly, it makes a social representation of the player towards the human's friends or other online competitors. Thirdly, it creates an AI representation of the player's tactics and skill level which can then appear in the games of other players, so that you do not need to be playing for your avatar to compete against your friends. Here again is the discourse in Kinect-brand promotional materials toward the erasure of boundaries between player and game, between human and nonhuman.

At the beginning of the game, the avatar creation process begins with a call to acknowledge the almost magical rendering of the human body within a game. The claim by the voice-over is that the console will use the player's "digital DNA" and create an avatar through the hardware's "technical wizardry". The avatar creation process makes use of bright lights and sound effects to highlight the magical nature of transformation from an unformed and unhuman collection of grey squares into an anthropomorphic form. Like the golem of Prague, unconscious matter is magically transmuted, except in this case the Kinect-artefact plays the role of the rabbi. The avatars and avatar creation process have been described as "surprisingly accurate"[267] but also as "a cartoon approximation" [268] and one reviewer claimed that the avatar "may or may not bear a likeness to you."[269] Several reviewers have criticised the logistics of the avatar creation process, which requires the player to kneel down next to the Kinect-artefact in what one reviewer called "a submissive stoop towards [the] TV".[270]

The original Wii and subsequent Nintendo consoles may have contributed to this, since the digital avatars or "Miis" represented players in several games. But again, the avatars in the Nintendo world are much simpler and cartoon-like. Nintendo's choice not to aim for true verisimilitude is in keeping with their minimalist and family-friendly approach to the industry. Similarly, Microsoft's choice to aim for relative accuracy in their portrayal of players represents their emphasis on high-performance machines with excellent graphics, a sensor that is not just a toy and their rebranding of Kinect-brand for adults. This hit-and-miss portrayal must therefore have been a significant disappointment for Microsoft, those objects that are in strong relation with Microsoft, and the Kinect-brand. The specific Kinect-artefact's frustration is much more situated in the moment of disconnect between Kinect-artefact and human user.

The idea of the avatar creation process is generally inoffensive and predictable, but as with any kind of biometric analysis it is important to seek out the restrictions of the technology. As a body-mimicking device for a game aiming to reproduce human ability, the game automatically creates a negative territory of bodies that do not conform to the requirements of the sensor and of the game. This is a territory containing individuals with low vision or hearing, mobility restrictions or without a certain normative body part such as an arm or ear. Any media form imposes certain restrictions on use including non-ergodic literature which at least requires "trivial" effort by the reader to perceive light and dark and to turn pages.[271] But the failure to encompass a broad range of players does seem contrary to Microsoft's stated intention to provide effortless play. The direction to kneel on the floor for avatar creation is, for example, an impossible feat for many individuals. It does not appear to support portrayal of individuals without two arms and legs, and most of the games can only be played while standing. On the support website for the Kinect-brand, Microsoft acknowledged that the Xbox One may not be accessible to everybody but argued that it is individual developers that must be "educate[d]".[272] The Kinect-artefact looks for certain ideal characteristics in humans, but every individual human presumably deviates in some way from these ideals. "The human" never presents itself to the Kinect-artefact.

The game causes a human body to position itself in a particular way. Part of this process might include moving furniture so that the Kinect-artefact can pick up more of the players' bodies. Once this process is completed, the individual is directed to either a sitting or standing position, then to move one's body parts to correspond to virtual stimuli, such as moving one's arms and making a grasping movement to hold virtual handholds during the rock-climbing game. On various online forums, some users reported that these actions are both possible and enjoyable with the ideal furniture and room size, but for many users without these luxuries their more strenuous moves were punished by a sensor that could not pick up all of their bodies. There was also some disconnect with the screen because the body is shaped both by what the Kinect-artefact detects (or is thought to detect) and by the screen, which may not be in complete accord. When shooting for a goal in soccer, the size of the screen may change the player's idea of where the goal and goalkeeper are located. Once the player is capable of grokking the system this problem disappears, but initially at least the body is shaped both by visual cues (screen) and by visual feedback (Kinect-artefact and screen). The concept of a "natural" user interface starts to become a bit tenuous at those moments.

There is little variation between champions, with all features changeable and found in menus. Two ages ("adult" and "junior") and two genders ("male" and "female") are available. Still, once the champion is created it is intended to be a representative of the player. Since the Kinect-artefact recognises individual users, each player can step into play and their specific champion is brought up at once. But what is more, the game apparently learns what kinds of behaviour an individual is likely to use at a given time. Examples include using power-ups at particular times and using particular moves to gain extra points. This is an additional level of complexity for the use of the avatars. *Wii Sports*, for example, restricts itself to assigning a numerical skill level to each avatar. *KSR* aims for a much stronger correlation between real player and virtual version of the player.

More abstracted from the game is the player's indirect behaviour after interacting with the game. *KSR* is designed to be a social game, so both the game and the player should benefit from engaging more players. The game becomes (ideally) more fun with competitors both in front of the same Kinect-artefact and connected via the internet. Therefore, the person who owns the Xbox One system is implicitly encouraged by the game to convince his or her friends to purchase their own Xbox One consoles. This is of benefit to the brand and the game, each of which wants to be successfully marketed and sold (see below). This could also be one of the motives for successful avatar creation, because it facilitates online play with friends. Champion scan artist Iain McFadzen claims:

> It really does put an extra edge on competition when the person you've just pulled of the rock wall in climbing, or barged in to a mine in wake racing, looks and plays like one of your friends or family. And, of course, as they're Champions, they're always available to play against, 24/7.[273]

Having a good representation of oneself in the game goes some way to breaking down the barriers of abstraction in online play, and being able to play with a person known in the real world can make games more enjoyable for the human (and is good for the Kinect-brand).

The only way for a human to investigate a Kinect-artefact is to look to the phenomena it produces and ultimately to look with human senses. An empirical approach must provide data for the task of interrogating the phenomenology of the Kinect-artefact. But the overmining or undermining of the objects can arise and must be assessed in each individual case. In this case the agency of the Kinect-artefact is closely tied to the agency of the game, but this is not always the case. Moment by moment the allegiance between the entities changes, but under most conditions they align. A Kinect-artefactish way of behaving is typically to receive data, transform it into useful information for the console and to provide feedback about its ability to sense certain objects to the human. We might refine this into a human-centric statement: the Kinect-artefact aims to appear to be a useful and effective NUI. The game behaves in a way that conveys the code for specific media formats. This can be refined into the human-centric statement that the game aims to provide good gameplay, graphics, sound and online play for the human player. Ultimately both the Kinect-artefact and the game interact with the Kinect-brand: aiming to impress the player, to be well-reviewed and to sell more copies of itself.

The conflict between Kinect-artefact and Kinect-brand is evident in *KSR* as in the "Waking up the Kinect" problem. The game wants to be liked and sold, being created by corporate interests and thereby being inextricably linked to them (rather as Latour's gun is inextricably and bidirectionally linked to the human – see Chapter 5). But this tenuous connection is easily broken by the many betrayals of the Kinect-brand and the Xbox One console more generally. The game's aim of good sales is obstructed by the failure of Kinect-artefacts to perform as promised. Specifically, the Kinect-artefact often fails to hear the player's spoken commands, it often cannot see the tops of players' arms while they are rock-climbing, and it does not always deliver an accurate image of the player during avatar creation. Kinect-brand programs of action are easily severed by the most immediate Kinect-artefact programs of actions. This is consistent with Harman's claim that objects are all equally real: the game disc, the ephemeral interaction between player and NUI, the corporation and the effect on the developer's reputation.

A similar incident is described by Ian Bogost in *Alien Phenomenology*. The game *E.T.: The Extra-Terrestrial* for Atari Video Computer System is discussed (a game so unpopular that the unsold copies were famously buried in a landfill). In a few pages, Bogost enumerates the multitudinous things that are meant by the name of this game.[274] He begins by describing the code, then the physical form of the individual cartridge and finally in the context of the 1983 videogames market crash: "not just a fictional alien botanist but a notion of extreme failure, of 'the worst game of all time': the famed dump of games in the Alamogordo landfill, the complex culture of greed and design constraint that led to it[...]"[275] He stresses that none of these examples is the "real" *E.T.* game. Similarly, there is no incarnation of *KSR* that is more real than any other (a problem discussed further in Chapter 4). But in both these games, technical and corporate failure are implicit and thereby reveal inner truths about the game and the console that are inadequately concealed. Typically, the game's goal is allied with that of the manufacturer, but the inner essence of the game might make itself amenable to interruptions by other elements such as the wall in one's living room that makes it impossible to get one's entire body into the Kinect-artefact's field of view, or the background noise that makes one's voice unintelligible to the Kinect-artefact, or the availability of internet review websites (in strong relation with the Kinect-brand) that make these problems known to the rest of the world. The corporation and the game ceased to remain on the same side from the moment the game was released to the public. And while their goals may be momentarily the same their relations are always defined by their inner real qualities, not by the allegiance. Technological failures are ready examples of bidirectional asymmetry in interactions.

Under even more unusual circumstances we may interrogate the alien phenomenology of these objects much more fully. For example, the effects of fire or freezing temperatures reveal other kinds of agency in these objects, such as the tendency of plastic components to melt before metal ones, or the tendency of a game disc to shatter when smashed with a rock. For Harman, blasting the console into space is a similarly valid investigation of object relations and allows the creation of novel sensual objects between the game and other entities. While the human experiences a far broader set of sensual objects by using the game and console as intended by the manufacturer, this is not necessarily true of the Kinect-artefact for which picking up human gestures is repetitive and is only a tiny portion of its potential. The Kinect-artefact is not reducible to its experiences, just as we are not reduced by the Kinect-artefact, which literally takes our image and leaves us as we were.

Alien phenomenology of the Kinect-artefact

The value of OOO lies in its strong theoretical robustness. It may lack the whimsical irreverence that seems so inextricable from the most successful ANT texts (such as in the paper by Jim Johnson, alias Bruno Latour, meditating on the incredible and seldom recognised power of the door hinge[276]), but it is a worthy successor for a scholarly milieu that is unsurprised by the agency of scallops and laboratory equipment. The tension of materiality and immateriality that surrounds posthuman discourse cries out for a philosophy of technology that can account for ephemerality and durability in human action, and OOO provides this with all the rigour and forethought that continental philosophy can offer. What OOO lacks is a grounded empirical framework that accentuates these strengths while contributing to a universal grammar for those who seek to replicate or extend other people's work. OOO and NUIs, in other words, have similar problems: a lack of widespread exposure; vehement opposition from groups that find more tried-and-true techniques more sensible and practical; a limited number of examples that herald greater things to come; and competing impulses toward either consolidation or innovation.

To begin a grounded empirical investigation is to experience a loosening of the strings that connect the strength and elegance of Harman's OOO from the object under observation. We are entering a world of alien affect and significance with limited language and tools. As Bogost claims, the only way to understand an object's internal existence and modes of perception is through analogy: "[t]he subjective nature of experience makes the unit operation of one of its perceptions amount always to a caricature in which the one is drawn in the distorted impression of the other."[277] Suitably performed, an anthropomorphic statement can accentuate the alien nature of objects such as the Kinect-artefact. One might refer to a Kinect-artefact's "desire" as a means of highlighting the strangeness of machine agency. The frequent use of anthropomorphisms by Microsoft engineers and marketing executives may fall into this category, although they must of course remain suspect because of their tendency to make products sound more "smart" than they really are. The aptness of any statement is, of course, subjective. The analogy must be sound or the anthropomorphism is suspect. In the case of the Kinect-brand, the promotion of the sensor made frequent references to the eyes of the Kinect-artefact, which as has been described led to inevitable disappointment from users.

The most fundamental Kinect-artefactish action is to take in information about its immediate world (while facing one direction). This is generally true unless the Kinect-artefact is broken or switched off, in which case its most noticeable relations might be with furniture and the ambient temperature of the room. These are all consequences of the real qualities of the real object, which exists only in the phenomenal world of the Kinect-artefact. To ask about the real object and its qualities, in Harman's sense, is to want to *be* the object. The observer can only ever have access to the sensual object, the one that appears in our impression of the object. As discussed in Chapter 1, for Harman the sensual object is the object that appears when one object forms a relation with another object. The sensual object makes it possible for one object to relate to another (since real objects cannot directly touch one another). The sensual object has particular sensual qualities that we humans are capable of perceiving. But this does not mean that there is no connection between the real object and the sensual one. The sensual object acquires its qualities from the real object. Harman escapes idealism by insisting upon this correspondence.

The connection between object and object is, as discussed in Chapter 1, a matter of vicarious causation for Harman. Consequently, the interaction between human and Kinect-artefact is bidirectional and of a different nature in each direction. Each object is taking something completely different from the other. The human anthropomorphises the Kinect-artefact and, in a manner of speaking, the Kinect-artefact Kinect-artefactises the human. That is, the figure of the human is converted into a form that the Kinect-artefact is capable of experiencing. When playing *KSR*, the human impresses itself upon the Kinect-artefact in the form of returning photons, infrared radiation, soundwaves, and light. These relations do not take anything from the real human underneath. Similarly, the human understands the Kinect-artefact in terms of its ability to see, its thought processes, and its menacing presence in the room.

Desire in the Kinect-artefact

It is easy to point to agency in the Kinect-artefact, but can we speak of desire? Desire surely suggests a gradation of needs that derive from the object's qualities. These range from fundamental necessities (such as electricity and connection to a console) which allow it to operate, to more minor wants (such as the correct angle of the camera and for the player to be properly positioned) which allow it to operate in an optimally Kinect-artefactish way. A good illumination of this gradation may be found in the attempts of the "Waking up the Kinect" investigators to derive an understanding of an alien entity through reporting on their various sensual relations with their Kinect-artefacts. The investigations are attempts to glean more information from the sensual object and to make assumptions about the real object. The human users, denied access to a reliable account of the Kinect-artefacts' relations with the world, initiated a program of research into the sensors' "ears" and documented their attempts online. These attempts are a speculative program which exhaustively lists all possible interactions that a Kinect-artefact could have with its environment. It tries to position the human viewer within the Kinect-artefact. It tries to understand what the Kinect-artefact wants.

The Kinect-artefact's world was thoroughly described through this program and some important insights were acquired. The investigators discovered that Kinect-artefact desire is variable and analogue, even if its code is not. What works for one unit may not work for another unit, despite the patently obvious fact that each unit ought to be identical. Each Kinect-artefact has a different history and a different environment, similar to the way in which human identical twins become different due to their environments. Furthermore, each unit has certain needs that are more or less important than others. It might be difficult to stoop down in front of the television to allow the Kinect-artefact to scan one's facial features, while the activation of a voice command might be as easy as opening one's mouth at the right time. Some actions are needed, some are desired and some are so easy they might be attributed to gestural excess. But there is an additional understanding to derive from this incident, and that is about what the Kinect-brand wants. The refusal of Microsoft to acknowledge the issue is a very telling part of the Kinect-brand's identity, which is one of intuitive use and seamless entertainment integration. The Kinect-artefact's needs and experiences are ludicrously removed from those of the Kinect-brand.

It may be argued that the idea of the brand is surely just the description in human terms of the object; the brand is an invention to bridge the gap between the objective world of objects and human thought. It is a desperate attempt to comprehend individual objects in an environment where thousands of identical units are shipped en masse. Bearing this in mind, is it right to have devoted so much time to discussing it? The Kinect-brand is a human thing, not a Kinect-artefact thing. This is perhaps another, grander, anthropomorphism: the allocation of human-like qualities not just to one object but to thousands. The brand comes to have human qualities (hope, disappointment, stubbornness, defeat). But it is also a matter of scale. The Kinect-artefact and Kinect-brand are reflected upon in a huge number of human ways: in pictures, on websites, in videos, and in webcomics. Each of these entities is a siphoning of the Kinect-artefact's real qualities into human media formats. There are very few entities with which Kinect-artefacts have a stronger connection.

In distinguishing between the Kinect-brand and the Kinect-artefact it is revealed that their needs come into conflict with each other. We have learned from this that the tiniest interference in the relation between human and Kinect-artefact can be catastrophic for the Kinect-brand. The tendency of the microphone of the Kinect-artefact to respond better to higher voices was immediately cause for anger in a large group of cis male gamers, the market which the Kinect-brand craved access to so ardently. In conceiving of a machine ethics this distinction would be relevant. As noted in the introduction to this book, desire is necessary for an object to make ethical decisions. There must be an opportunity for an object to decide between an action that is "right" and an action that is in its own self-interest. In the case of the Kinect-brand, if we accept that the object has something similar to desire, then we could say that its self-interested desire to insinuate Kinect-artefacts into gamers' homes comes into conflict with what is seen by many as the "right" things to do, such as ensuring that human voices are equally easy to understand regardless of pitch or accent. As for the Kinect-artefact, its desire to listen all the time for the words "Xbox on" may come into conflict with its user's right to privacy.

Death of the Kinect

The "death" of a non-living object can only be discussed when using a rather playful (if morbid) anthropomorphic eye. Clearly things that are not alive cannot die, but we use the word often to describe a wide range of non-living objects. In the novel *Slaughterhouse Five*, Kurt Vonnegut brings it to the reader's attention by commenting "So it goes" after each mention of death.[278] This statement emphasises the death of humans and also the ways that we connote death in the English language, such as the death of the novel. The death of both the Kinect-artefact and the Kinect-brand are phenomena that force us to ask how much an object needs to change before it is no longer the same object.

Not all Kinect-artefact experiences revolve around the human. In no other place is this more evident than in the discarded Kinect-artefact, thrown onto the vast electronic waste repositories that now litter our planet. Electronic waste may be considered a "hyperobject", to use Timothy Morton's term.[279] The hyperobject ("hyper" relative to us) is something that has such a powerful influence in either space or time that it denies the subject-object relationship with humans from which we are accustomed to framing the world. Morton is chiefly concerned with the most currently pertinent hyperobject: global warming. But he also describes the universal qualities of hyperobjects. One of Morton's criteria for hyperobjects is that they are "nonlocal"; that is, any immediate, present version of the object is not the object itself. So, the Kinect-artefact that is present in the moment is not the hyperobject. The Kinect-artefact's period of existence will likely far outstrip our own. It has already been brought about through the siphoning of ancient hydrocarbons through an oil rig (literally, not just metaphorically as in Harman's *Circus Philosophicus*) the removal of ore from the earth and the refinement of minerals into glass. Even if these processes do not strictly speaking constitute the life history of the Kinect-artefact (since objects are irreducible), the future existence of a Kinect-artefact is certain to also be highly eventful. We cannot be certain of the future of this and other electronic artefacts, but at present it seems likely that they will be abandoned to decay over thousands of years. For almost all of this time the main interactions for the Kinect-artefact will be with objects other than individual humans. If we consider time as a continuum then the Kinect-artefact occupies a field that stretches over this long period. Any human contact occurs in a tiny portion of this continuum, and much of the object is very distant.

As has also been mentioned, an even more abstracted "death" of the Kinect-brand is said to be occurring at the present moment. The Kinect-artefact's decay and redistribution of objects in an unhuman world will likely be agonisingly slow (from a human perspective), but the Kinect-brand can be injured quite easily. The Kinect-brand is dying because it is no longer being sold with each Xbox One (and may even require a special adaptor),[280] which means that fewer games will be made and fewer people will talk about the brand online.[281] At 2019's E3 press conference, Microsoft made only fleeting reference to the sensor. As of 2018 the Kinect-artefact has almost entirely disappeared from gaming contexts. Online forums and websites almost universally agree that the Kinect-brand is dying or dead. But what exactly is meant by the "death" when it applies to a brand?

Possibly the brand dies when games stop being made and the mainstream media is no longer interested in it.[282] Now that not all, and perhaps the majority of, human users interact with an Xbox One without a Kinect-artefact, developers have reverted to making relatively few games for use with the Kinect-artefact. This has caused the marginalisation of the Kinect-artefact for Xbox One by Microsoft. But there is also the Kinect-artefact for Microsoft Windows, which may see more positive results. In this case, the brand specific to the console may die but its structure, sensors and software may live on. Or perhaps the death of the brand is no longer in the hands of the corporation. The flourishing hobby of retrogaming or old-school gaming attests to this possibility. Obsolete or discontinued interfaces may have a rebirth in this context, in which the collection and enjoyment of classic and unusual gaming objects is paramount. Objects can live on in this form until they are ready to begin their gradual physical decay on one of humanity's electronic waste dumps. But can the brand be conserved in this way as well? The Kinect-brand is less a corporate trademark and more an idea in a broader sense, which in this case does not need to die when it ceases to be supported by Microsoft, much in the same way that *Pac-Man* or the Nintendo Game Boy are not dead ideas. As long as there is a community of Kinect-artefact users the Kinect-brand may be said to have survived.

This chapter was an experiment in distinguishing between two distinct objects: the Kinect-artefact and the Kinect-brand. If there are two different objects, then there are two different sensory worlds which are very different in nature, and each requires a different methodological approach. Sensation for a machine is immediately fairly evident to a human, but the nature of it can be hard to pin down. We have witnessed the struggle of the human user to find its phenomenal boundaries, which given its collective and collaborative nature is probably one of the best ways of asking the Kinect-artefact about its experience. What is the ethical consequence of this investigation? The alien world of the Kinect-artefact is far removed from those of humans and higher animals, those beings that rank so highly in our collective estimation. Should we be concerned for the well-being of the Kinect-artefact (or even the Kinect-brand), perhaps carefully maintaining working units as we would conserve an endangered species? Or perhaps we should make sure always to perform a calibration before playing? Clearly the possession of "eyes", "ears" and a "brain" are typically not sufficient to incite a user to care for their electronic device in such a way. The death of the Kinect-artefact or Kinect-brand can hardly be considered cruel. If humans do not care about these things, then why bother with alien phenomenology at all?

At the start of this book I asked to what extent the sensation of anthropomorphic machines is relevant to their cultural role, and alien phenomenology emphasises the sensation of nonhumans. In the case of the Kinect-artefact there are certainly practical reasons why we should care about the machine's sensation of the world. We should strive to understand any object that threatens our privacy. And we should study emerging technologies thoroughly before they become integral to our lives. But the real advantage of alien phenomenology is its capacity to appreciate objects like the Kinect-artefact and the Kinect-brand for their unusual way of being in the world, so different from our own. We know that the study of the Kinect-brand is of interest, because understanding the nature of brands is an essential quality of people who work in many academic and professional roles. We know that the approach is of interest for the users of the Kinect-artefacts because some of them performed their own informal alien phenomenology in a collaborative online attempt to form stronger relations with their consoles. It was an analysis that Microsoft, with its talk of brains and eyes, could not perform without damaging the potential of the Kinect-brand. A far greater potential for human-avatar-console is afforded by the Kinect-artefact. The seeming independence of the Kinect-brand from the Kinect-artefact is remarkable, even for videogames hardware. They are a remarkable model for charting the distribution of human and machine agency in two complex and contentious objects under a deceptively simple umbrella term: *Kinect*.

Chapter 3
Grey Walter's Tortoises and an Introduction to the Sensation and Experience of Robots

> Building machines included a simultaneous building of milieus for the machines.
>
> Jussi Parikka

The goal of this latter part of the book is to provide a rigorous critique of what has already been written about the perspective and experience of robots in post- and nonhuman discourses and how they may relate to alien phenomenology and OOO. Robots defy neat classification, with only the qualities of automated movement and human design being universal to all. Robots vary enormously. Like the Kinect-artefact, some are designed to be more specialised, and all their sensory systems are honed to a particular task (or so it seems to a human user). Some can perform a variety of tasks with a multitude of sensory systems that are in use at different times, as is the case with Voyager 2. Obviously, sometimes, their degree of anthropomorphism is far greater even than the Kinect-artefact, which as we have seen is often described in terms of its "sight" and "hearing". A robot that assembles cars is even more difficult to compare to a human, despite possessing "arms" and sensory capabilities such as proprioception. Nevertheless, the metaphor of human-like qualities and behaviours is impressed upon the human mind. This chapter will problematise these notions of fixedness and anthropomorphism with relation to the simple example of robot tortoises.

The primitive robot tortoises *Machina speculatrix* were designed in the late 1940s and early 1950s by cyberneticist W. Grey Walter. They were around the size of a toaster and were featured prominently in Walter's writing. They are emblematic of the early years of cybernetics and robotics; a physical manifestation of the application of human psychology to machines that was so important at the time. The tortoises possessed two types of sensors to guide them around their environment: light sensors and touch sensors. They offer a charming opportunity to talk about relations between different materials. But while it is interesting to speculate about the material relations that the tortoises had or have, the actual objects of the research presented here are multifarious. Since objects are irreducible, every part of the tortoise is itself an object (wheel, motor, sensor, etc.). This chapter will also discuss related objects like drawings of the robots and texts in which they featured. *Machina speculatrix* is also an incorporeal object that has its own qualities and relations, much as *Homo sapiens* or humanity has its own qualities and relations. None of these objects is more genuine than any other; they all exist, they are just different.

This chapter will consider how we can speak of sensation and experience in robots, given that our understanding of these things is rooted in human experience. This book posed three research questions in the introduction, and one most relevant to this chapter is "What does sensation mean for machines?" Given the vast variety of robots, what are we saying when we ask about the experience of robots? We must acknowledge the huge range of functionalities and experiences possible for robots. It is rather like studying yeast and the hummingbird separately as distinct species with distinct experiences while holding to the idea that both are alive. Performing alien phenomenology means learning about the experience of individual objects. This chapter presents and critiques two different ideas with reference to the robot tortoises. The first, Umwelt-theory, privileges the importance of analytical investigation of an entity's worldview; the other, theories of the concept of mind, emphasises the heterogeneity of mind or consciousness. They are both ways of questioning the experience of an entity, but Umwelt-theory begins with the entity itself and builds up the concept of experience from the ground, while theories of the mind begin with the question of experience before asking about the mechanics of an individual entity's way of knowing the world. Both ideas are relevant to anthropomorphic machines, but they are used here to contrast each other and provide extremes of robot analysis.

When we attempt to theorise robot phenomenology we are discussing a range of experiences, but the language in which our description is couched is ultimately motivated by the particular ontology of the robot used. Having endeavoured to illustrate the phenomenological consequences of OOO in the previous chapter, this chapter will make available alternative theories of nonhumans and technology in particular. The domain of robots presents a varied terrain in which light might be shed by the use of the different ideas, much as the ethologist of hummingbirds must become acquainted with a different set of ideas to understand the behaviour of yeasts.

Umwelten and cybernetics

The comparison of animal with robot, if done with caution, provides a starting point for discussing their particular sensation and experience of the world, and alien sensation more generally. The work of Baltic ethologist Jakob von Uexküll (1864-1944) is a case in point. Uexküll's Umwelt-theory is heavily influenced by Kant,[283] and Uexküll was certainly no anti-correlationist. The Umwelt allows us to see the link between the subject and the environment as composed of codes that intertwine to determine one another. Uexküll's work, which is part zoological and part philosophical, was for many decades obscure, but pieces such as the monograph *A Foray into the Worlds of Animals and Humans: with a Theory of Meaning* (1934) are now quite well-known in Western philosophy and are seen to communicate complex thought about the world of animals, particularly those that do not share human sensory equipment.[284] The key to this is Uexküll's emphasis on sensory equipment as constituting animal perception, although Uexküll avoided anthropomorphic concepts when speaking of the Umwelten of other species.[285] Uexküll's work has also been discussed in Jussi Parikka's 2010 book *Insect Media* and Uexküll's curious way of accounting for his observations has begun to enter the vernacular of post- and nonhuman thought. Uexküll has also been referred to in descriptions of animal affect written by Heidegger, Deleuze and Guattari, Canguilhem, and Merleau-Ponty. Umwelt-theory's identification of the body as the site of subjectivity predates Merleau-Ponty, and Deleuze and Guattari's interpretation of Uexküll highlights the semiotic component of this body-centred subjectivity.[286]

As an inquisitive scholar of biology and of philosophy, Uexküll's writing is often explicitly preoccupied with the major philosophy of biology questions of his time. Uexküll was not a Darwinist, and he did not adhere to a mechanistic view of living things. This has led some to cast him as a vitalist.[287] But as Kaveli Kull points out, Uexküll considered himself to be something in between these two extremes, a "machinalist".[288] It is important to bear this point in mind, because it is indeed tempting, when encountering Uexküll's views on the status of non-living things, to argue that some sort of vitalist philosophy must have been present. In some ways it is also surprising to note that he did not hold a mechanistic view of life given that some of his ideas appear to be precursors to cybernetic ideas. But the emphasis in his texts is on the semiotic relationship between subject and environment rather than being confined to how effect follows cause.[289] Uexküll's main interest was in how living things sense the world, and how to create a language for talking about those sensory experiences. The concept for which Uexküll is best-known is the Umwelt (sometimes translated as "environment" or "milieu", but better equated here to "subjective, meaningful world"). Umwelten are comprised of all the elements of an environment which the living thing is capable of sensing; in Uexküll's words it is the "phenomenal world or self-world of the animal".[290] Uexküll's writing is replete with musical analogies that capture both the essence of his subjective model for biology and his conception of a "plan" in nature that is responsible for the seemingly perfect harmony between the Umwelten of different organisms.[291] The Umwelt concept, in the form of the science *Umweltforschung* (Umwelt-research), was developed by Uexküll over the course of his life and has been revived by his son Thure von Uexküll, Thomas Sebeok and others, particularly in his native Estonia but also more widely.[292] It is a major contributor to the founding of the field of biosemiotics. Since Uexküll's work was largely lost to the non-German-speaking world for several decades it is only in more recent years that its importance has been identified and his position as a founder of the field of biosemiotics, particularly after Thomas Sebeok and Jean Umiker-Sebeok's 1992 publication *The Semiotic Web 1991*.[293] Biosemiotics is, broadly, the use of sign theory to describe biological phenomena, predominantly that of animals (zoosemiotics) but increasingly also that of plants (phytosemiotics) and other organisms.[294] Jesper Hoffmeyer even synthesises Umwelt-theory and the evolution of the "virtual" genetic code.[295]

The Umwelt is the world of experience for any given animal, containing things that can be sensed and excluding things that cannot be sensed. Everything that can be sensed enters the Umwelt as a sign, and the Umwelt contains the mechanism through which the sign is sensed, perceived, and responded to. It is the perceiver-effector of the animal's experiential world and it is described in semiotic language. This might mean that the Umwelt relates only to the parts of the light spectrum that the animal can perceive, or only to certain chemicals that the animal is capable of perceiving with its olfactory sensors. The model organism used by Uexküll in his books *Umwelt and Innenwelt of the Tick*[296] and *A Foray into the Worlds of Animals and Humans*[297] is the female tick during its penultimate life stage, in which its experiences are truly alien and can assist us to dissociate from our anthropocentric view. The tick suspends itself from a leaf or blade of grass, waiting, sometimes for years, for the only stimulus it is honed to sense and respond to: the butyric acid released by mammals. When it senses this chemical, the tick drops with the hope of landing on the mammal's body. This accomplished, the tick utilises its two other senses, permitting the sensation of temperature and hair typology, to find a suitable place to pierce the skin and feed on the mammal's blood. Once it has drunk its fill, it lays its eggs and dies. These three senses represent the totality of the tick's universe at that stage in its life cycle. The three senses form the tick's Umwelt, or the semiotic "bubble",[298] which is the tick's entire world. Two different organisms of different species may share the same surroundings, but they do not share the same Umwelt. The Umwelt concept does not require us to find analogies to compare the tick's experience with our own. For example, it is tempting to describe the tick's "smelling" of the approaching mammal, an inaccurate anthropomorphism that equates the use of the arachnid's alien olfactory organs to our own sense of smell which is structurally different and evolved separately.[299] This side-steps the problematic impulse that may be found in much of the literature around OOO: the very great temptation to go too far in our description of nonhuman worlds and thus to humanise them and rob them of their great power of alienation. The power of alienation in biosemiotics is increased when we look at the vast variety of Umwelten that exist in the animal kingdom (and possibly beyond) and compare them to the Umwelten of humans. The human Umwelt is inextricably linked to the things that help us to survive. We cannot see very small things or things that are very far away, we can only detect a small range of chemicals with our olfactory and gustatory systems, and we only perceive certain, very specific, types of radiation. As Uexküll says, sensation is like a "garment"[300] that wraps around us, revealing only certain parts of all possible experiences of the world. If we were to apply this idea to technological artefacts as well as living things, we might see a similarity between the limitations of the tick's Umwelt and that of the Kinect-artefact when it is in its instant-on state. As far as we know, the Kinect-artefact's Umwelt at that time is geared towards the sensation of sounds, allowing the Kinect-artefact to identify the "Xbox on" signifier that will prompt it to switch on the Xbox One. As for *Machina speculatrix*, we have the advantage of knowing a bit more about their structures than those of the blackboxed Kinect-artefacts. Walter's writing gives us some clues as to the nature of the tortoises' Umwelten. But despite their unhuman qualities their cybernetic context encourages the drawing of similarities between the robots and humans. How fortunate we would be if Uexküll had lived to use his skilful and poetic analytical powers to help us dissociate from an anthropocentric view of robots as well as that of ticks.

Uexküll, in his many surviving texts, does relate the Umwelt concept to technology on more than one occasion. But according to Parikka, because he was living and writing in the early twentieth century, his idea of machines "meant clocks, factories and blindly repeated processes"[301] and his personal philosophical statements betray mistrust of the idea that living things could operate solely according to the same principles as non-living things. Again, he is sometimes accused of vitalism for this reason. A machine, such as a telescope, could certainly be used as an extension to a human's Umwelt, but the telescope itself does not have an Umwelt according to Uexküll.[302] More recent writers have taken the Umwelt concept to a place Uexküll could not reach, with the aid of technologies that perhaps do more to spark the phenomenological imagination. We now have machines that are capable of activities that resemble perception to a far greater extent than the telescope. As mentioned above, humans lack certain elements in their Umwelten. But we now have visual technologies that transcend scale and distance by transposing the sensation of a machine with something that we can interpret, such as the translation of electrical signals into visual or audio media. Machines can "smell" the environment and relate their sensory experience of a particular chemical in numerical form so that we can understand it. Individual machines can pick up far more types of radiation than humans. And Voyager 2 can signal to Earth information about the ice giants of our solar system without our needing to go there ourselves. Claus Emmeche has asked the question "Does a robot have an Umwelt?".[303] Even primitive robots have "simple functional circles of semiotic processes of sign-interpretation and sign-action. Why should they not have Umwelten?"[304]

The Umwelt as applied to technology bears some similarities to the concept of affordance. In James J. Gibson's book *The Ecological Approach to Visual Perception*, the affordances of an animal's environment and how they can be changed to be more suitable is discussed.[305] The affordances of the environment are the aspects of the environment that are relevant to the animal. For humans, if a surface is "horizontal, flat, extended, and rigid" then it affords support (as well as being "fall-off-able").[306] Affordance is a different concept to the Umwelt, because affordance implies not just the availability of objects to be sensed, but the perception of usability and adaptability by a human or nonhuman: "[Affordance] is both physical and psychical, yet neither."[307] Umwelten do not encompass the value of the object for the human or nonhuman, but only contain the representation of sensory information. Don Norman extends the concept of affordances to the design of objects by humans, arguing that we adapt objects to offer new or different affordances in particular contexts.[308] A large part of the designer's efforts must go into signifying the presence of features in designed objects because features can exist without being perceived by the (human) user, and only become affordances if the user is compelled through context, knowledge and capability to use the feature.[309]

Affordances are apparent in all different kinds of humans and nonhumans, and the scope of research into objects is wider than with the Umwelt. But because the affordance involves the perception of value of the object, there is an assumption that the human or nonhuman must have some capacity for experiencing its environment (Norman extends this capacity to humans, animals, and robots[310]). In contrast, the Umwelt invites the question of whether sensation must be tied to experience. It is not clear whether all nonhumans *experience* their sensory worlds in the same way, merely that there *are* sensory worlds. That is, the existence of nonhuman sensation is not dependent on possession of a mind, consciousness, qualia or any kind of experience. Thomas Sebeok argues "there can be no semiosis without interpretability – surely life's cardinal propensity",[311] so does a robot need to be able to interpret its environment to have an Umwelt? As will be seen later in this chapter, the mind of a robot is a highly contestable entity, with the extremes of functionalism and panpsychism resulting in the attribution of vastly different kinds of internal experiences for non-living things, and even living nonhumans. It will likely be some time before we can really answer the question "Does a robot have an Umwelt?", if indeed it ever happens. Then again, perhaps the nature of interpretability is vague enough for us to apply it to nonhumans regardless of whether they possess minds or consciousness. This book's interpretation of OOO certainly does not require objects to have minds to have sensations or experiences.

There is a convergence between OOO and Uexküll's research program, though they approach similar conclusions from different roots. OOO also rejects a purely positivist research program in favour of a richer world of relations, and an interest in the diversity of sensory capabilities in objects. In *Theoretical Biology* Uexküll was critical of physics for privileging the objective over the subjective. For him biology was the domain of the subjective:

> Physical theory tries to convince the plain man that the world he sees is full of subjective illusions, and that the only real world is much poorer, since it consists of one vast, perceptual whirl of atoms controlled by causality alone. On the other hand, the biologist tries to make the plain man realise that he sees far too little, and that the real world is much richer than he suspects, because around each living being an appearance-

world of its own lies spread, which, in its main features, resembles his world, but nevertheless displays so much variation therefrom that he may dedicate his whole life to the study of these other worlds without ever seeing the end of his task.³¹²

Uexküll's emphasis on a "richer world" could easily come from a text on OOO. The world of nonhumans is alien and constituted of alien sensory capabilities. But in OOO subjectivity and perception are less emphasised. Subjectivity and perception both imply the presence of a mind or mind-like quality, and tend to reinforce the subject/object dualism. As will be shown later in the chapter, some OOO thinkers have embraced or at least considered panpsychism. But Harman's body of work does not *require* panpsychism. The notion of experience in biosemiotics is the result of the perception of a sign, not just the change in the nature of the real object. For example, the tick's Umwelt contains only three capacities for sensing the world. If I throw a rock at the tick causing it to fall off its perch, then that does not enter into the tick's Umwelt. But in OOO terms a sensual object is created between the rock and the tick. Not everything that happens to the body of the tick is in the Umwelt, even if it affects the tick significantly. The Umwelt only relates to signs equipped to filter through the tick's sensory organs and then interpreted, even if this occurs in a very alien way, which is why biosemioticians tend to be reluctant to extend the theory to non-living things.

As will be seen in the next section, Walter's robot tortoises are comparable to a tick due to the apparent poverty of their sensory worlds. They possess sensors that allow them to detect light and touch, and they have internal sign processes such as the transmission of electricity that is interpreted by components of their bodies, such as the motor or battery. Like the tick, and like humans, they do not pick up everything in their immediate surroundings. If the crux of this problem lay solely in questioning whether a robot may possess an Umwelt due to its human origins and electronic sensory-perceptive systems, it could be dismissed quite quickly. But can we say that the robot perceives and interprets its world, and how would we know? The case in point here is the robot tortoises. After describing the robots' existences and how they worked, this chapter will renew this enquiry into their internal experiences.

Robot tortoises

W. Grey Walter created several tortoises, two of which were named Elsie and Elmer.[313] Elsie and Elmer were first exhibited to the press at the end of 1949, and more advanced tortoises were built in 1951.[314] Robotics was an extension of Walter's work on electroencephalography (EEG) and the brain, and began as a project in his spare time.[315] Walter was a member of the Ratio Club in London, a pioneering group for cybernetics.[316] Walter gave his new "species" the "mock-biological"[317] name *Machina speculatrix* and observed that from their simple interactions they could develop quite complex behaviour.

Machina speculatrix tortoises recall the tradition of building ceremonial or amusing automata. The building of automata was once a common practice both in Europe and parts of Asia. Automata are mechanical toys, powered by clockwork, that commonly move in a repetitive way, and are very often built in human shapes or the shapes of animals. Jacques de Vaucanson's gold-plated duck is a famous example.[318] There was a minor craze for automata in the sixteenth century amongst the aristocracy in parts of Europe. But interestingly the practice was curtailed by church authorities, who eventually came to believe that the automata represented a heretical attempt to replicate the work of God, namely the giving of life to a non-living thing. This is an oft-mentioned comparison: Bruce Mazlish's analysis of Frankenstein declares that "[m]an, the evil scientist, has taken God's place."[319] Lois Kuznets argues that this restriction may have been a manifestation of the engrained Judeo-Christian taboo against the production of graven images.[320] Jessica Riskin describes an episode of iconoclasm in which medieval devotional automata were destroyed, the iconoclasts claiming that "[m]achinery [...] could not represent divinity other than deceitfully."[321] In Japan, this practice was not as restricted, however. The building of automata was and is a job reserved for the most respected masters of craft. *Karakuri*, as they are known, play a prominent role in Japanese culture, notably participating in festivals and other events. Possibly the traditional animistic nature of Japanese culture and religion means that *karakuri* do not seem as heretical as they did in Europe. Objects of great spiritual importance, such as a sword, hold a soul or *kami*, so *karakuri* and humanoid robots are not the only non-living objects with a soul.[322] Walter followed in this tradition when he created these fancifully-named and spectacular robots, but the cyberneticists of the time were *deliberately* blurring the line between humans and nonhumans.

From a human's perspective, Walter's tortoises constituted a significant leap forward in that they were capable of semi-autonomous movement and apparently quite complex behaviour. They were dissimilar to the automata in that they possessed significantly more independence of movement and sensors that seemed analogous to human sense organs. They were driven by an electric motor. A tortoise would wander around until it hit something, at which point a contact switch would be activated and it would move away in a different direction. Walter wrote a brief technical explanation in *Scientific American*, interestingly attributing the change in direction to "an elementary form of memory".[323] It was intended to replicate a negative feedback loop. There was also a photocell on the front of the machine that would cause it to move towards light, including their lit-up hutches where their batteries would recharge. This was intended to replicate a positive feedback loop. However, at a certain level of brightness the tortoises would move away from the light, so that they would get close to the light but not too close. As the battery depleted, the sensitivity to the light source became greater and the tortoises could detect the hutch from farther away, but if the battery was completely charged the tortoises would move away from the hutch and explore other areas. The tortoises could therefore return to their illuminated hutches when it was time for them to recharge, feeling their way around obstacles in their paths, and making them leave the hutch and "explore" when the battery was charged.[324] Walter restricted Elsie and Elmer to two feedback loops to "discover what degree of complexity of behaviour and independence could be achieved with the smallest number of possible interconnections."[325] Elsie and Elmer's Umwelten, if they can be said to possess such things, consist of light and touch at the very least. In a OOO sense, there are other elements to the tortoise's world as well – Elsie's alien sensation may also extend to the texture of the floor under her wheels and changes in temperature that affect her metal parts. But it is particularly important to note that although Elsie's Umwelt may contain light, it is not the same as the light in human Umwelten. Elsie's light is constructed by the material of her light sensor.

The tortoises also had small lights which were originally placed to indicate whether the motor was running. When the tortoise was moving, it emitted light, but did not emit light when stationary. The effect was such that when a mirror was placed in front of a tortoise, it would seem to gaze at its own light in the reflection and make small movements back and forth "like a clumsy Narcissus."[326] If two were deployed at once, they would bump against each other, attracted to the other's light but repelled by the touch of their shell. In Margaret Boden's words, they seemed to have an "unseemly fascination" for women's legs, possibly because women wore nylon stockings which reflected light to which the tortoises were attracted.[327] To watch the BBC newsreel of *M. speculatrix* is to be astonished that such simple rules could result in what appears to be complex behaviour (although it is likely that the tortoises were manipulated and made to produce more precise movements for the purposes of the film).[328] The movements of the tortoises were also tracked using long-exposure photography. Walter affixed candles to the tortoises' shells in darkness, and the effect was to produce an image of long bright lines showing where the candle (and thus the tortoise) had moved.

Walter and others were fascinated by the functional anthropomorphism they could engineer in quite simple machines. In Walter's writing about the tortoises it is clear that he believed significant insights could be gained by observing these artificial beings. *M. speculatrix* was not built to perceive light and touch as a human would. But there was a set of impulses reminiscent of life, like those observable in a creeping vine. A creeping vine senses and responds to light by angling its leaves to catch the most sunshine, and it responds to touch by building structures that wrap and bind. For the cyberneticists, this performative sensory-motor response was the essence of living nervous systems. In the case of the tortoise gazing at itself in the mirror, Walter declared that "were it an animal a biologist would be justified in attributing to it a capacity for self-recognition."[329] Of course, the tortoises were not capable of self-recognition, possessing just one photocell and only two functional elements. But cyberneticists sought to create responses to stimuli, not to recreate organic cognition.

The functional anthropomorphism of the robot tortoises is in stark contrast with the visual anthropomorphism of automata and *karakuri*, in which the visual element is much more important. An automaton's *Umwelt* is located much closer to the surface. The puppet or statue's sensation of turning wheels is a different kind of analogy from that of Elsie's sensation of light, and certainly not analogous to human vision. It is instead the impression left upon the artefact of the human prime mover in the form of the gradual reduction of tension in the mainspring and methodical ticking of clockwork. A tick moves in response to a smell and an automaton moves in response to human intervention. No wonder they seem blasphemous when they are activated from within and only from human intervention. On the other hand, the lack of functional eyes and ears necessitates the addition of other human-like features like faces, a feature which Elsie and Elmer lack.

What kinds of experiences can we attribute to robot tortoises? For a few years they occupied a unique position in the boundary conflicts arising between the categories of human and nonhuman brought on by cybernetics. The nature of robot sensation and action were a reaction to and encouragement for more blurring of human/machine boundaries. This has seriously affected the way that we consider their existence. It certainly affected Walter's scientific and popular portrayal of the robots, and may be noted in his description of them exploring, resting and considering themselves in a mirror. With the knowledge that the cyberneticists had limited success in creating an artificial mind in the 1950s and 60s, it is easy to look back at these accounts and smile at Walter's attribution of complex behaviour to such simple robots. But perhaps all such anthropomorphic language is doomed to failure. The robots performed the functions required by Walter to exhibit his thoughts about the mind (although by all accounts they frequently broke down, and were, therefore, disobedient). They did spend periods of time positioned in front of the mirror. They did move erratically around the space as though exploring. And during that time they formed sensual objects with the different real objects around them. They related and, in small ways, changed, such as when they changed direction.

Sadly, relatively little is known of the robots' design. Knowing more about their structure and relations would assist in the use of alien phenomenology, but we must make do with what information has survived. We know that the 1949 batch were pieced together with spare materials: old alarm clocks and war surplus materials.[330] Walter's 1953 book *The Living Brain* is more a guide for how to build one's own tortoise rather than a scientific study of the robots Walter had created.[331] Unfortunately Elsie and Elmer are not thought to have survived, so we cannot be certain of their structure or the relations they might have formed.[332] There are very few images or records of intermediate stages of their construction. This is a serious methodological problem for alien phenomenology. But in an object-oriented description, the robots are not simply identified by their components. Elsie is an object, the battery is an object, the light sensor is an object, the image above of her inner mechanisms is an object. No object is defined by another. As discussed below, we may embrace the flexibility of the name "Elsie" and acknowledge the changeable nature of a robot. Its material structure changes, its qualities change and its relations with other objects changes. Elsie's tendency to break down, capacity for alteration and eventual loss are facets of her Elsieness. And given the sketchy accounts of her short period in the spotlight, these might be some of the only aspects of her that are accessible to modern humans, machines and texts.

Tortoises and time

The study of the tortoises depicts the presence in time of all objects, no matter their immediate appearance of solidity or impermanence. The weirdness of the tortoises brings these matters to the foreground and makes them a good model for explaining Harman's concept of time in object-oriented philosophy. Firstly, the Elsieness described above smears a changing abstraction across several decades while also maintaining her own integrity. In other words, the tortoises maintain their existence as objects despite constantly changing: from the alteration of component parts to the slow degradation of iron when it comes in contact with oxygen. How can we account metaphysically for these objects remaining themselves while simultaneously becoming something else?[333] For Harman the Elsie with her shell on is still the same object as the Elsie with her shell off, merely being encrusted with outer changes. Yet for Latour the irreducibility of objects means that a new object must arise with every change, even ones as small as the light on the front of the tortoise switching on and off in front of a mirror. As Latour says in *Irreductions*, "[t]ime is the distant consequences of actors as they each seek to create a fait accompli on their own behalf that cannot be reversed [...] In this way time passes."[334] Latour's actants change constantly, shifting allegiances and switching relations. Actants do not lose or gain extra distinguishing characteristics over time, rather they become different actants as they change.[335] For Latour, Elsie with her shell on is one actant, Elsie with her shell off is another, albeit closely related, actant.

In *The Quadruple Object* Harman offers an alternative perspective. His inspiration comes from Husserl's phenomenology (with several significant caveats). In Harman's interpretation of Husserl, the stability of objects over time is depicted as the solid core of the object surrounded by a changing flux of "adumbrations".[336] An adumbration is a particular perspective of an object at a given time, such as the view of a water tower as one walks around it. As discussed in Chapter 1, objects are withdrawn from relations and it is sensual objects that facilitate interaction between objects. So when we gaze at a water tower, we are experiencing the sensual object that is formed between us and the water tower. If we walk around the water tower later in the day in a different mood, it will appear differently to us. But we are unlikely to mistake it for a different object. The sensual object formed between us and the tower remains the same even though its sensual qualities change. We can identify an object despite the "gems, glitter and confetti of extraneous detail".[337] The tension between the sensual object and its encrustation of shifting qualities is one of Harman's main lessons from Husserl, and he labels this tension "time". While a sensual object remains the same for us, we encounter sensual qualities as "intermittent changes."[338] Unlike for Latour, real objects and sensual objects remain the same over time, but the encrustations shift and change. In Harman's theory, there is an inner Elsieness that remains whether her shell is on or off.

Similarly to the distinction in the previous chapter between artefact and brand, we must make a distinction here between the relations formed by Elsie in the moment and our relationship with Elsie so many decades later. Our typical relationship with the tortoises can only be by proxy. Their motion is recorded in human media such as in newsreel footage and

Walter's sketched drawings in *Scientific American*, as well as presumably finding a place within the brains of those people who witnessed them first- or second-hand. Personally, I have had no direct relations with the tortoises. I have seen images and read texts about them, including texts by Walter himself. But it is the images and texts with which I have a relationship, not the tortoises. The tortoises only exist in relationship with objects when that relationship is bidirectional. When we read a paper on the tortoises, there is no bidirectional relationship with the tortoises, only with the text. There is a paperish mediation. There are the long-exposure photographs of burning candles positioned on the robots' shells; the effect on the cybernetic community of the early 1950s; and the renewed academic and cultural interest in these and other early robots: these are all consequences of the tortoises' existence, but they are not the same as Walter's observation of the tortoise nor of the adhesion of the candle to the shell. The tortoises' worlds were characterised by alien interactions. Many nonhuman objects were engaged with the robots. The wheels of the robots were in a relationship with the floor beneath them, and the kind of interaction that occurred was doubtless different depending on the particular surface. They were subject to the earth's gravitational pull, the slow oxidation of iron and the qualities of different replacement parts. Brenton Malin points to a similar example:

> For object-oriented thinkers, sidewalks are not merely signifiers of certain social relations. They interact with shoe rubber, dog paws, bicycle tires, and other objects in – quite literally – concrete ways that give them an important agency and power.[339]

Perhaps a *Machina speculatrix* shell is at this moment home to an industrious hermit crab, or perhaps it is at the bottom of a pile of garbage. In these cases, the shell's relations with other objects are not recorded, and do not appear in history books, but figure in different kinds of struggles with ocean currents and gravity unrecorded by humans. These may seem like mundane and trivial considerations, but to fail to point them out would constitute the erasure of the validity of the access of other objects.

In addition to the actions of other agents in the early 1950s it is important to speculate about the robots' enduring experiences that are relatively distant from human interaction. The tortoises are lost, destroyed or relegated to museums: an unsurprising end to these once remarkable machines that have since been hugely surpassed by more modern inventions. They are representative of the kinds of histories that technological artefacts possess. Contemporary media theory is beginning to express an interest in the experiences of objects beyond their interaction with humans, as was discussed in Chapter 2 with regard to hyperobjects. The hyperobject is "hyper" relative to human lifetimes and individual human agencies. In addition, media archaeological approaches are increasingly preoccupied with what happens to technological artefacts after they are abandoned by humans. Siegfried Zielinski's book *Deep Time of the Media* explores the existence of lost and forgotten moments in the history of media that were only precipitated far in the future.[340] In a lecture, Jussi Parikka reframed this concept to theorise the material history of technological artefacts: from the bringing forth of natural resources to build their components to their eventual decay or repurposing.[341] Since we do not know the fate of several of the robots it would be presumptuous to claim knowledge about their futures, and even the future of the tortoises that have survived is uncertain, although there is a good chance that they will survive, encrusted with changes, longer than we will.

These weird objects have held a certain fascination since the 1950s, representing a key example of cybernetic thinking. More modern writers are interested in the way that the robots changed the scientific project at the time and their long-term status as paradigmatic objects. The robots have endured for humans well beyond their time in the public eye. Yet they are also lost objects, ill-defined from the outset and so frequently invoked that they begin to lose their definition. In their timeless state, and in their inability to be present, they expose the effect of time on objects.

Sensation, experience and consciousness

Let us compare the framing of the tick with that of the robot tortoises provided here. In Uexküll's conception, the tick sits, stationary, possibly for many years, with only very limited means of sensing the world. The three senses that enable it to locate and feed from mammals, such as the sensation of butyric acid, are the only relations that the tick has with the world. The light of the sun bouncing off its body, the contact between moisture and legs, the movement of the grass in the wind: these do not count, because they do not enter into the tick's Umwelt. As a corollary, we must decide which of the robot's relations qualify as "senses", which qualify as parts of the Umwelt, and which do not. (Uexküll would strongly object to this, but some more modern writers would not.) Light and touch, surely, qualify. But the contact between wheel and ground is probably more problematic. Who can say which elements are being "sensed" and which are other kinds of relations? Which part of the robot is even doing the sensing or experiencing? Emmeche explains this difficulty by pointing out that there is neither analogy between human and robot nervous systems nor evolutionary homology between the two systems.[342] If the structure is completely different, how can the experience be similar? Winfried Nöth claims that the symbolic representation portion of a robot's experience might correspond to the concept of the Innenwelt, more closely allied to the animal's interior experience or "I".[343] This may present a model for conceptualising artificial intelligence in the language of Uexküll.

We are surrounded by machines that appear to sense things. If we reduce the components of the robotic Umwelt to signs, then the issue may not entirely be resolved. Asking a robot questions about its sensation may require resorting to desperate measures. If a robot tortoise is suddenly doused with water, the water acts as a sign to the robot's electronic components, disrupting its functions and causing it to shut down. If a robot tortoise is exposed to extreme heat, those temperature signs are interpreted by the atoms in the robot's metallic components as an instruction for their atoms to move more quickly, causing an expansion in the metal and eventually melting. If a robot tortoise moves across an uneven floor, it may be jostled. The change in the texture of the floor may be sensed by the wheel which can no longer travel along its original trajectory. The wheel may, in its way, perceive and interpret that sign and respond by changing the position of the wheel, which is made possible by the weight of the robot and the power of the motor. At no point is any kind of decision made, but the robot responds to the conditions in a similar way to the growth of a seed when its cells respond to gravity, some cells moving upward to become the stem and others plunging downwards to become the roots. If biosemiotics can be applied to any living thing, including things without any kind of neural system, then what exactly do we mean by sensation and perception? Can we distinguish sign-relations from the universal relations that underpin Harman's theory of objects? Is the robot's experience anything like biological subjectivity? These are questions that relate strongly, although not exclusively, to questions of mind and consciousness.

The Umwelt is conceptually distinct from the mind or consciousness. As Emmeche says, the mind is a much broader concept.[344] The Umwelt may be related to sensation and perception, but it does not depend upon the entity possessing/performing things as elaborate as consciousness, cognition or thought. So, in considering the Umwelt of an object, we encounter a difficult and ancient question: is sensation only done by minds? Our human sensation of the world seems to be strongly connected to our minds. So, one might reason that alien sensation also requires some kind of mind or consciousness. Uexküll's tick example, though carefully pared down and limited primarily to the function of the semiotic bubble rather than to reflections on what the tick feels, can only evoke an impression of an entity that processes stimuli much the way we do – as experience. Do we have to interpret that experience as a manifestation of mind or consciousness in the arachnid's tiny brain? A "yes" response to that question is surely at the root of Uexküll's purported vitalism and more contemporary biochauvinism. Humans struggle to fathom experience without consciousness. Consequently, Voyager 2 is no more capable of experience than stones or water. Of course, this leaves unanswered the question of where the cut-off should be between those entities we deem to have experience and those that do not have experience. If the tick qualifies, do its larvae qualify? Does a sponge? Or a mushroom? This is a very limited view of experience.

We have two choices if we seek to allow nonhuman or non-living things to experience the world. Firstly, we can accept that experience is limited to conscious entities but expand our concept of consciousness to incorporate any system of sufficient complexity. We can protest that consciousness is simply an emergent quality of matter that has reached sufficient complexity; our supercomputers and robots, past, contemporary or imaginary, may be said to possess that indefinable capacity for experience because they, like us, have consciousness. In so doing, we need not confront the experience of nonhumans who encounter the world without a mind. This is a popular (if vague) view. Of particular note here are those transhumanists, such as Ray Kurzweil, who believe that so great is the resemblance between natural and artificial minds that each of us may one day be able to transfer our consciousness between them.[345] A mind is something that happens when matter is organised a particular way and cannot exist without that matter. Most of this kind of speculation is derived from functionalism (and identity theory, which functionalism has mostly superseded).[346] Functionalism holds that mind is a quality that arises from the organisation of physical matter. It is a materialist view of the mind that is both a cause and result of the extensive research that has been done on the mind during the past century. In part because of identity theory and functionalism, psychology and neuroscience are influenced by the view that the physical laws that apply to the rest of the body also apply to the brain (and consequently to the mind), which helps to explain how there can be any interaction between them. Without the belief that brain and body are subject to the same mechanised processes, so the argument goes, it would be impossible to explain how the brain senses pain in the finger, or controls a muscle in the calf. There would have to be some sort of other intermediary, which we have not yet been able to find, otherwise the mind and body are impossibly separate. One solution is that of the medieval Islamic occasionalists to which Harman refers.[347] But the decline of spiritual explanations for the mind (for example, the existence of the soul) has led to increased belief in a monist, physicalist model. We will be able to identify the presence of a mind in a supercomputer because it will feature qualities similar to human minds. Thus we establish ourselves as the possessors of an ideal model of mind, to which other mind-havers will ultimately be compared.

The second possible option is to reject the idea of a hierarchy of mind. One way of doing this is to argue, as have many thinkers since ancient times, that the mind must be a quality of all matter, or that all matter must have the ability to produce a mind-like quality.[348] The word "hierarchy" is used deliberately here in provocation of the concept of flat ontology, and indeed some prominent scholars associated with OOO and the nonhuman turn do take this idea seriously, as discussed below. The idea that mind is a quality of all matter can be called panpsychism, and refers to the attribution of a mind, consciousness or a soul to all objects, including those objects we would usually think of as inanimate, such as robots, rocks and electrons. The word comes from the Ancient Greek roots *pan* (all) and *psyche* (mind or soul), and David Skrbina argues that it is "almost certainly the most ancient conception of the psyche" in the form of animism and polytheism.[349] Thomas Nagel may have described this notion best, by claiming that it comes from a few simple premises "each of which is more plausible than its denial, though not perhaps more plausible than the denial of panpsychism."[350] Panpsychism may seem incredible now, but as Skrbina points out many prominent philosophers (focussing for the moment only on the Western tradition) have subscribed to some variation on panpsychism culminating in considerable popularity in the nineteenth century, before it became less popular with the rise of logical positivism in the twentieth century.[351] Explanations of exactly how panpsychism works are rarely consistent between scholars, and even the language used is highly problematic. When we talk about panpsychism, are we speaking of consciousness, sentience, thought, qualia, the soul, experience, a psychological being, or the mind? Panpsychism may not be best thought of as a theory in and of itself, but that it is instead a commitment that may accompany radically different ideas. Consequently, it may be located in the work of thinkers as diverse as Thales, Spinoza, Leibniz, Whitehead and William James; with both monists and dualists, materialists and idealists.

Contemporary panpsychism follows this pattern in being a result of various philosophical positions. One such position is a continued adherence to Whitehead's process philosophy, but there are also instances in which panpsychism seems to arise as a modern solution to modern problems. David Chalmers' information theory invokes a view of panpsychism in exploration of his functionalism with, in John Searle's words, its "tacked on"[352] theory of consciousness. In his book *The Conscious Mind* Chalmers argues that consciousness (or experience) is a basic element of all matter by virtue of its capability to carry information.[353] This would appear to be an answer to Hobbes' point that sensation surely cannot exist without some sort of memory, and a reiteration of William James' characterisation of waves upon a beach as a sort of memory.[354] The waves leave markers in the sand of where they have been. In Chalmers' parlance, the gestures of Elsie and Elmer create information, which could be the foundation of their consciousness. In humans, that information could simply be changes in neuronal structure, hormone levels and so forth. In the robot tortoises, the information could be the tiny indentations made in the floor as they move, the movement of charged particles in the batteries and the slow degradation of iron. To be clear, Chalmers is not saying that information makes a thing conscious, but he does remain "agnostic"[355] about it, and about panpsychism in general. In Searle's scathing review, he calls this an "absurd view"[356]:

> There is not the slightest reason to adopt panpsychism, the view that everything in the universe is conscious. Consciousness is above all a biological phenomenon and is as restricted in its biology as the secretion of bile or the digestion of carbohydrates. Of all the absurd results in Chalmers's book, panpsychism is the most absurd and provides us with a clue that something is radically wrong with the thesis that implies it.[357]

Panpsychism is certainly not a mainstream view, but it does seem to be increasingly tolerated due to the influence of the computer as a paradigmatic case in philosophy. It is less and less common to posit a universe in which only organic life is capable of consciousness. Our greater understanding of the relationship between the brain and the mind and the exponential growth in computer speed and complexity forces us to consider the consciousness of inorganic entities. This has led some, but by no means a majority, to panpsychism.

Of particular importance here is what Chalmers calls the "winking out" problem: the question of which entities can be said to be conscious if we accept that consciousness is only something that happens to humans, or higher mammals, or all living organisms, or anything of sufficient complexity be it carbon or silicon-based. For any of these premises to be true, there must be a certain point when descending through the hierarchy at which consciousness winks out. Perhaps a bat is conscious but a bird is not, or perhaps a fish is conscious but a lamprey is not. According to functionalism, consciousness is a physical phenomenon that will be perfectly explicable once we have gleaned sufficient knowledge about the brain. But if that is true, how can one creature possess mind and qualia and consciousness, while another creature that is very closely related not possess these things? Where is the cut-off point, and how can it ever be anything but arbitrary? This leads naturally to the position that consciousness, as a human-like phenomenon, might not necessarily wink out as we descend the hierarchy of living things, but might change seamlessly. Consequently, a tick will still have something that we have to call "consciousness" or something similar, even though it looks totally different from a human's experience.

Why should we stop at the boundary between living and non-living things? A highly speculative panpsychist section of Chalmers's book is called "What is it like to be a thermostat?", an echo of Thomas Nagel ("What is it like to be a bat?"[358]) in which he describes the thermostat as "an information-processing system that is almost maximally simple."[359] All that is required for consciousness (although probably not thought) is a system of "causal interaction"[360], in which an external stimulus causes a change of state. This means that electrons meet the criteria, and Chalmers's speculative panpsychism is almost total. Of course, this kind of panpsychism does not include fictional or incorporeal objects.

Panpsychism is very difficult to intuitively accept if one is asked to begin from the principle that comets and sunflower seeds are sentient or conscious. But if, as Chalmers appears to have done, you begin with advanced computers and the hard problem of consciousness, the notion is more palatable. Supercomputers must have a mind if the dominant functionalist model is correct, since they behave as though they have a mind. Minds are most easily identifiable in machines that seem to be more like us, even if they are structurally very different. This is precisely the reason why this chapter juxtaposes panpsychism with a very simple robot. It is possible to draw a pleasing comparison between a panpsychist position and Grey Walter's enthusiasm for his robotic creations. A naïve view would see little difference between the attribution of mind or experience to the tortoises and Walter's rhapsodising over his "clumsy narcissus" and its powers of self-recognition. Panpsychism actually reveals a much more heterogeneous concept of mind: one in which the activation of switches and purring of motors is proof enough in itself of experience, rather than merely the vehicle by which the robot cons its way into consciousness by mimicking human proclivities.

If we extend mind, or at least subjectivity, to all things, then to some extent the novelty of the Umwelt is lost. An implication of suggesting that all information is a characteristic of mind is that all interactions are significant, and perhaps even that they may be represented in semiotic terms. In that scenario, the feedback loop may become all but interchangeable with Uexküll's concept of the function-circle. The correction of a robot's wheels as it moves across the uneven floor are as significant as its perception of light because the Umwelt encompasses the whole of the robot and every piece of information is part of its mind. Similarly, the motion of red blood cells throughout a human's circulatory system is a part of its Umwelt. Our brains might not be aware of the movement, but our arteries, tissues and plasma sense and experience it. In adopting panpsychism, it would be easy for the Umwelt to become irrelevant. But there is great value in considering the two ideas at the same time. The strength of the concept of the Umwelt lies in its power to force us to confront alien worlds. In the latter example, it might be possible to accept the presence of a mind in not just the whole human but in the individual blood cell itself. An inquiry into the subjective world of this blood cell could yield fascinating results. As for the robot, the imposition of a division of parts and wholes, the sort that we employ when studying living things, might be contrary to the nature of non-living things. The wheel is connected to and strongly related to the robot, but that does not mean that the two components bear the same relation as that between the human and the blood cell. The wheel-mind might be totally different from the robot-mind, or they might be aspects of a whole. Panpsychism may seem at first to be in conflict with the idea of the Umwelt, particularly with Uexküll's description, but an adoption of panpsychism might only require the adaptation of the theory to other kinds of sign-relations. As a bare minimum, it does not seem unreasonable that some adaptation be made so that robots can be said to have Umwelten, irrespective of the debate over whether they have minds.

Panexperientialism is sometimes suggested as a more moderate view of panpsychism; less an argument about consciousness than about "experience". Gregg Rosenberger claims that it is "milder"[361] because it does not require anything like cognition or mentality.[362] Again, it is presented as a solution to the mind-body problem. The concept of a "mind" is highly anthropocentric. A statement about the nature of an entity's mind can be an ontological statement about the nature of its experience and consequently the question of subjectivity, whereas panexperientialism leaves the interpretation a little more open. The term was coined by David Ray Griffin. He describes the interpretation as "Whiteheadian-Hartshornean".[363] Griffin says that "psyche" implies too high a level of experience for more elementary units of nature.[364] As Griffin puts it, Whitehead's philosophy begins with the idea of "prehension". Prehension is something that we share with all other individuals, regardless of whether they have sensory organs.[365] Every individual has experience through prehension. But not all objects are the same. Some are "compound individuals", which have a higher level of experience and that might have consciousness; others are "aggregational societies" which do not have higher levels of experience (Griffin gives the examples of rocks and telephones, and Elsie would also qualify).[366] Every individual prehends, but not all prehensions lead to consciousness. This happens in human beings too. Not everything that a human prehends makes its way into the conscious mind, only "those that have been prehended with the greatest intensity will survive to this phase; the rest are blocked from becoming conscious."[367]

Aggregational societies do not have experience according to Griffin; they are not "individuals". Elsie does not have experience, but the atomic and subatomic particles that constitute her do have experience. In the words of Cornel du Toit, the "resulting position can be called process philosophy's version of 'panexperientialism', which is applicable to all individuals but not to all things whatsoever."[368] Curiously, even though the "psyche" is dropped from the word, inanimate objects as such are still excluded. Prehension resembles the model for a concept of "sensation" presented in this book but is quite distinct. Throughout this book uses the word "sensation" to align the concept with Harman's ontology (and also as something of a provocation of our tendency to anthropomorphise). But I would not use the word "prehension" because of the rejection of experience in all kinds of objects found in "process panexperientialism". Fundamental to this objection is that all sorts of different things are objects: atoms, metallic strips, motors, wheels, and robots. The part/whole model of OOO is entirely different from that found in Whitehead. The word "experience" in "panexperientialism" is not the same as the usage in this book. In this book's terms, experience is evident in a response to sensation; the collision of real qualities with sensual objects. Panexperientialism is primarily presented as a solution to the mind-body problem and nonhumans are not usually the focus.

Harman's version of panpsychism is predicated on the irreducibility of the mind. The relation of a mind to an object outside the mind has the same ontological status as fire burning cotton.[369] Panpsychism is not required for flat ontology, or vice versa, but there is some association between the two despite the initial misgivings Harman revealed in *Guerrilla Metaphysics*, in which he rejects "human knowledge" as being of "pivotal importance in the universe".[370] In *The Quadruple Object* Harman describes a softening of his ideas towards panpsychism, even if he is not prepared to adopt the label. He uses the term "polypsychism"[371] instead, which is intended to highlight his point that a perceiving entity is a relating entity, and some entities are "dormant"[372] and free of relations. Harman explains that these objects may exist in principle, or perhaps that an object may have relations at one time and no relations at another (although he does not provide an example). Therefore, while experience can be attributed to most entities it cannot be attributed to all.

Harman is not primarily interested in minds. But the idea of minds has historically been such a feature of the distinction between humans and nonhumans that it is relevant to OOO. Consider the sensation of a bowl of soup by a human mind, which the human experiences as hot. But in addition the human experiences many things that do not prompt any changes in the mind such as the ambient temperature and the evaporation of water from the surface of the skin. Vesicles in an axon terminal release serotonin into the synapse and it is taken up by the dendrite of another neuron. The waves wash over the beach. Mario completes a level and finds that the princess is in another castle. Each is an example of sensation of an object by an object, just as a human senses soup and soup senses the human. They are yet more examples of real objects that create sensual objects to allow interaction, but they create a change in the real objects. They produce encrustations of sensual qualities that change the way that the object is able to form relations. At a very basic level, the object is affected by this change, and must register it. In other words, an alien kind of experience is taking place.

The robot tortoises don't need a human-like mind to experience things either. The mind may be just another kind of object. Thoughts are sensual objects that appear when a relation is formed between the soup and the mind. When we limit experience to things that have a mind, we are framing everything through a human concept. Even if we concede that a robot has a very different mind than a human, it is still the word *mind* that we use. The word is misleading, steeped in history and particularly in the human/nonhuman (and mind/body) dualism that OOO minimises. But we don't need to be anthropocentric in our use of the word. The nature of the mind does not, in principle, matter, because it is as indisputably an object as unicorns and phlogiston. As Steven Shaviro points out, a common criticism of panpsychism is that it uncritically assigns human-like mind elements to nonhumans.[373] This, Shaviro says, begs the question because it betrays an understanding of thought, value and experience that is inherently anthropocentric. Consequently, the emphasis of this book is on experiences rather than minds, although not in a panexperientialist sense. This is because if experience is universal to all objects it must apply to incorporeal objects as well, not to mention inanimate objects. The experience of a synapse, wave or videogame character does not need to be a deviant version of a human experience. Experiences are just as alien as the kinds of sensation that define an object. In a robot, the response to sensory input is evidence of experience. It signifies that something changes inside the object. This is not a panpsychist argument that depends upon the presence or absence of a mind connected to sense organs. The tortoise does not use its sensors to detect when it is lifted up in the air by a human hand, but it still senses and experiences it.

This chapter has focussed on one of the guiding questions of this book: what does sensation mean for machines? More specifically, does using the word "sensation" in a strict, non-metaphorical sense require us to believe that machines experience the world in any meaningful way? This chapter has argued that in the context of Umwelt-theory, it is not necessary to anthropomorphise experience, but it is necessary to recognise some kind of semiotic exchange between nonhuman and environment. It is natural to be reluctant to extend a quality of mind to robots as simple as *Machina speculatrix*, but it is not necessary to do so to speculate about their experience of the world. It is difficult but possible to conceive of non-anthropomorphic minds or experiences, with or without a panpsychist view, and even without a panexperientialist view. The Umwelt is conceptually distinct from the concept of a mind and yields a very different account of how an object experiences its world. Umwelten give an account of a sensory world that need not be anything like our own, despite taking place in the same environment. The concept of a mind relates to the experience of those sensations. For the anthrodecentric approach presented in this book both Umwelten and minds may be seen in at least all physically embodied objects. Yet both the Umwelt and the mind have historically been restricted to certain classes of objects because of the assumption that both depend on the presence of subjectivity. As OOO rejects the primacy of subjectivity, the concepts of Umwelten and minds can be better deployed as explanations of the elements of Harman's metaphysics (and must be, if object-oriented ontologists can acknowledge them to exist at all). The Umwelt of an object is the impression of the environment upon an object; it is the field in which sensual objects are formed and maintained despite the flickering encrustations of their qualities. Notice that in this view the Umwelt must incorporate entirely alien sensations such as the impression of wheels upon a floor. The Umwelt is the sensory world. Meanwhile the mind, if the concept is made sufficiently broad, describes the way that the object encounters sensation. It describes a very fundamental kind of experience. It is nothing more than the sum total of the changes brought on by the sensation of other objects as those sensations change moment by moment. In this account, the Umwelt and the mind are not special objects relegated to the realm of living things for the sake of human vanity.

Anthropomorphism is a trap which must be avoided when speaking of Umwelten or minds. Objects that are more human-like do not have non-alien sensation and experience, whether they are parts of humans or models of humans. Synapses, waves and Mario are all alien. The point is to avoid a hierarchy of mind-having or Umwelt-having objects, beginning with humans at the top and negotiating the positions of various anthropomorphic objects,

objects characteristic of the Anthropocene or objects that typically inhabit human bodies on a spectrum. The point to bear in mind is that we and our sensation and experience are as alien to other objects as they are to us.

Chapter 4
SHRDLU and Bidirectional Postphenomenology

Postphenomenology differs from traditional phenomenology in that the structure of human perception and experience is changed by technologies. As Tom Sparrow puts it, phenomenology is almost by definition about humans and human lived experiences through the "principle of intentionality that sutures subject and object together", but it varies greatly from scholar to scholar.[374] Postphenomenology does not make the same radical claims as OOO but it does significantly depart from (what we might call) traditional phenomenology by emphasising the role of technology in affecting human experience and even being a part of human experience. Technologies are not merely mediators of human experience, but constitute our reality.[375] Don Ihde is the most prominent theorist in this field. For Ihde, human perception and other intentionalities extend through machines. Technology must not be considered abstractly, but in terms of human-technology relations.[376] Many of Ihde's examples are visual ones, such as the non-neutral effect of Galileo's telescope on stargazing.[377] Despite requiring a degree of trust in the medium to convey an image that expresses the world in terms humans can understand, the telescope nevertheless presents the human user with objects in the solar system that would otherwise be invisible, such as the moons of Jupiter. Indeed, in the case of modern telescopes, distant galaxies and phenomena outside the visible light spectrum may also be glimpsed. The scholarship that has arisen around this concept contributes to a growing appreciation of the role of technology in human phenomenology.

Ihde is not an OOOer and he is not overly concerned with deconstructing the categories of "human" and "machine". His postphenomenology explains an important and often neglected part of humanity's lived experience. But for the purposes of this book postphenomenology is typically far too one-sided. To only use Ihde to understand humanity's lived experience is to reject the OOO position of anti-correlationism. Therefore, this chapter will present, as far as is possible, strictly bidirectional and asymmetrical interpretations of Ihde's work. Where Ihde describes a human-technology-world relationship from a human position, this chapter will attempt to describe a similar one from a nonhuman object's perspective. This is an experiment in combining two approaches, but it is hoped that this bringing together of approaches yields a useful language and framework for describing the asymmetrical, bidirectional relations between human and technological artefact. This may bring us closer to the goal of conceptualising the way that machines sense and experience the world. Since postphenomenology is so robust in its treatment of human-world relations, by reversing the process we can make coherent statements about machine sensation.

The computer is a paradigmatic metaphor for human experience. We saw in the previous chapter how a technological artefact may affect and shape our line of inquiry about the human mind. This chapter focusses on a computer program named SHRDLU as it forms relations with its environment, an environment that includes human interlocutors. SHRDLU is (or was) a natural language processing (NLP) program designed by Terry Winograd in the late 1960s and early 70s. SHRDLU's world was entirely concerned with moving and answering questions about crudely illuminated block shapes on the system's monitor. The setting is deceptively simple, concealing a vast, complicated web of interrelated processes that make it possible for SHRDLU to "understand" English. Through the constructs of postphenomenological theory this chapter interprets the ways that SHRDLU extends or changes human actions and perception, and how these constructs might be reversed to describe the way that technological agents use humans.

What is (or was) SHRDLU?

The world of SHRDLU is deliberately tiny. It is an entity with a miniscule capacity for direct action in its environment. Like the tortoises of the previous chapter, it has a small set of ways of sensing the world, the most pertinent for most human analyses being the way that it senses humans. SHRDLU's only information about its human user comes to it through textual orders and interrogations, relayed through the keyboard of whatever system it happens to be occupying at that time (or possibly, these days, through voice-to-text applications). The human types instructions or questions, and SHRDLU responds with text and/or actions in its block world. Once that information is received, SHRDLU's capacity for action is very limited. It can only perform actions as it has been programmed to do. Apart from responding to the user's dialogue, its sole possible set of actions consist of moving around digital blocks with a digital robotic arm (the digital block world is visible to the user).[378] The blocks exist as code within the computer but are programmed to obey the laws of the physical universe, so for example, a block once dropped will fall onto whatever is below it. This is the microworld of SHRDLU, the epitome of "hacker" style artificial intelligence programs designed by Terry Winograd between 1968 and 1972, created to investigate natural language and computers. It has become a classic piece of AI programming, providing what was at the time a fresh perspective on the relationship between world and language. Like the tortoise, SHRDLU's simplicity is the key to its engaging theoretical implications; it becomes a model for greater things, providing clues to unlocking more universal insights. In a 1979 paper Hubert Dreyfus discusses SHRDLU along with other microworlds created in the late 1960s and 1970s, such as Adolfo Guzman's 1968 SEE program which was designed as a simple computer vision program.[379] Dreyfus is critical of these programs because he believes that human intelligence consists of "local elaborations of the whole" rather than isolated microworlds that assemble to form the everyday.[380]

Nevertheless, Dreyfus calls SHRDLU a "major achievement."[381] The program is far more complex than the interface suggests. Like any elegantly-designed user interface, it conceals the complexity of its underlying processes. The language use of programs like SHRDLU is one of the most studied and contentious areas of artificial intelligence research. Humans have a fascination with reproducing their forms in artificial entities, and especially in reproducing certain defining human qualities like language. The intelligence modelled by early systems like SHRDLU was limited to tightly constrained domains and therefore imitated only a tiny part of human intelligence. Playing chess and proving mathematical theorems were possible, but one could not enjoy a disinhibited conversation with a machine.[382] SHRDLU cannot converse with a human about anything other than its block world. This so-called "weak AI" may be contrasted with the "strong AI" theorised by science fiction writers and futurists, in which human intelligence is not merely simulated, but from which consciousness and self-awareness emerge.[383] Strong AI is not simply an improvement of processing power, but a replication of the linguistic and experiential qualities of the human brain, and supporting the argument that the brain is *like* a computer (as in functionalism).[384] The supposed need to observe understanding and explanation in machines is often seen to be a major problem in the development of AI. Creating and responding to signs is seen as a step above simple performative and adaptive systems. Human brains *understand*. They interpret the world through language and other signs and draw conclusions after the application of symbolic logic.

One of the earliest applications of computers to be seriously considered in AI circles was the translation of texts from one language to another. But in 1960 Yehoshua Bar-Hillel proclaimed that the high-quality translation programs sought after and expected were nowhere near being completed, and used the phrase "the box was in the pen" to illustrate his point.[385] He pointed out that no program at the time could translate the word "pen" as used in that sentence correctly, the way that an English-speaking human would be able to do immediately. The machine is in need of "extra-linguistic knowledge". Winograd argues that early machine translation failed because language consists of more than just the concepts of "dictionary" and "grammar"; language must be connected to an external world.[386] As he puts it, "the computer didn't know what it was talking about."[387] A conversation is part of each participant's ongoing attempt to gain information about their environments.

A successor to the translation programs, conversational programs were built between the mid-1960s and early 1970s. SHRDLU is one such conversational program. They were attempts to build systems that had a limited understanding of language. One famous example is ELIZA, a program developed by Joseph Weizenbaum at MIT between 1964 and 1966. Despite the limitations of the time, ELIZA was made to be a somewhat convincing conversationalist by running the DOCTOR script in which it was themed as a Rogerian psychiatrist talking to a patient. This is a clever move because the inevitable non sequiturs can be mistaken for insight in a psychiatrist identifying hidden links.[388] A Rogerian psychiatrist also often reflects back what the patient is saying to draw the patient out and get them to keep talking, which is something the program could handle.[389] Winograd points to this episode as a warning that, with clever tricks, a program can appear to have more insight than it does.[390] Its simple gestures to anthropomorphism also strongly imply the presence of a human-like intellect. ELIZA has some similarities with contemporary virtual therapist Ellie, funded by DARPA to aid in the diagnosis of mental illness amongst veterans.[391] In some ways, Ellie's goals are the same as DOCTOR's, since it is important for her to connect with "patients", who must be encouraged to see past the computer interface to the connection with another entity. This is aided by Ellie's advanced natural language processing as well as her ability to pick up on bodily cues and her own use of body language. But the important difference here is that Ellie has an aim beyond fooling human interlocutors with her use of language. She is not a purely theoretical project, and she exists to diagnose rather than to interpret. There has already been some literature discussing the advantages of virtual therapists: for example, people seem to be more willing to discuss sensitive or embarrassing topics with a computer.[392] But there are also risks, as in the use of therapeutic anthropomorphic toys in children with autism, such as distress when the agent is taken away or hacked with malicious intent.[393]

It has been argued that these capabilities are beyond the artificial brains that exist in the world to date. John Searle, the critic of David Chalmers mentioned in the previous chapter, demonstrates this line of thinking in the Chinese Room thought experiment.[394] Searle's is a well-known argument associated with the feelings of disappointment of the 1980s that came from the failure of AI to live up to expectations.[395] A non-Chinese speaker is put inside a room. From outside the room, a Chinese speaker writes a question and puts it through a slot into the room. The person inside the room has access to a book containing many different Chinese phrases and the appropriate response to give. The person can copy down the response and return it through the slot. The result, to the Chinese speaker, appears to be comprehension by the computer and a salient response. However, the entire exchange takes place without the person inside the room understanding the question or answer. Searle argues that this is how a computer appears to return intelligent answers despite no comprehension taking place. However, Ray Kurzweil disagrees and claims that the Chinese Room model is a construct that reveals what occurs in simple single-neuron systems.[396] In a system like the human brain, great complexity and volume of data transforms the performative neural response to a supposedly semiotic one through processes of emergence, and there is no reason why, given sufficient complexity, something like thought might not occur in a computer.

This question is pertinent to the creation of modern NLP agents such as Watson, the IBM computer reportedly capable of writing poetry. The depth of human skill and wisdom required to write good poetry is considerable, so it is easy to see why comparable creative output from a computer would lead the casual observer to ascribe it to human-like intelligence. In 2015, the IBM Watson team capitalised upon this in an advertisement starring music legend Bob Dylan.[397] Watson and Dylan discuss the themes in Dylan's poetry, including love and the passage of time. Watson notes that although it can read 800 million pages per second (and can clearly parse and interpret poetry), it has never known love. Watson's natural language capabilities are extremely advanced and are informed by the input of vast quantities of text, so that it is able to answer sophisticated questions on a huge number of topics.[398] In one sense its knowledge of love is doubtless extensive. But it is not grounded in the world with respect to that notably human emotion. Its sensation and experience of love is alien. Given Watson's incredible linguistic skill, however, it may well be perfectly capable of writing poetry that mimics Dylan's characteristic lyrics. Whether it understands them the way a human would is another question, as discussed in the previous chapter.

Language is considered one of the defining qualities of human nature, and it is only to be expected that we would try to reproduce it in computers, if only to see if it is possible. But the practical possibilities for such an endeavour are also enormous. One need only look to the contemporary examples of Amazon's Alexa and Microsoft's Cortana to see the vast cultural value attached to communicating with machines "naturally", and consequently the economic value thereof. As discussed in Chapter 2, this goal of interacting "naturally" with a machine is problematic but appealing. Natural language communication with computers is also a common trope in science fiction and is commonly a quality of advanced user interfaces, such as the computer character Samantha in the film *Her* (played by Scarlett Johansson).[399] In this film, the ability of the computer user to communicate through natural language with his computer leads to an intense intimate relationship. The computer is anthropomorphised in other ways as well, but her presence is located in her ability to speak; it is the facilitator of her sensation of the human.

SHRDLU can be seen as an earlier relative of today's more advanced NLP programs, hampered by its inability to access information about anything other than its block world and system. The world in which SHRDLU operates is necessarily small so that it can understand everything that goes on within it. Winograd argues that "success at understanding language depends on a deep knowledge of the subject being discussed."[400] Deb Roy calls this "grounding", the process by which an agent relates beliefs to external physical worlds.[401] In this way the computer need not have the world mediated by anchor-less human language but can have a direct experience of the objects it is discussing. It would have been impractical to attempt to make the computer relate to objects in the real world; this was far beyond the technology at the time. And for OOO purposes it is not important whether the world of blocks was "real" or not (see below) – it still constitutes a collection of objects, and we can look on SHRDLU in OOO terms as an early attempt to extend the sensation-poor environments of computers. Winograd had a reason for this attempt that was both practical and philosophical, and it resulted in a system that had a wealth of information available to it. More information means, correspondingly, greater agency within its world; that is, the world of blocks. The information about its world was not just coming to it through the intermediary of language, but from its sensation and understanding of the block world.

Speaking to SHRDLU is like speaking to a toddler that is only just learning about its world. In his 1972 paper in *Cognitive Psychology*, Winograd provides a detailed technical explanation of SHRDLU's means of understanding and answering questions, along with a transcript of a human conversation with it.[402] This is a typical interaction:

Person: PICK UP A BIG RED BLOCK.

Computer: OK. (does it)

When the robot arm completes an action, that action is reflected on a visual representation. In this case the robot arm descends from the top of the screen and picks up a red block meeting a pre-determined definition of "big". As will be shown, this simple exchange requires what was at the time a very advanced understanding of language and logic. As Douglas Hofstadter says, SHRDLU is a "mathematical ignoramus" which has only been taught to count up to ten.[403] But its simplicity is, in fact, the key to its solidity as a paradigmatic example of a particular phase in the history of AI.

Winograd, whose experience in computer science led him to become a philosopher, is of the opinion that previous attempts at NLP failed because there was a lack of connection between the language the system is being asked to understand and a real world that it can understand and interact with. With this in mind, he built SHRDLU as not just a teletext exchange between user and computer but incorporating the simulated "blocks world" with physical laws and rules for an aspect of the system to interact with. To the human user, the teletext screen is presented in order for them to provide instructions to or ask questions of SHRDLU, and for SHRDLU to respond. But there is also a screen showing a simple world of blocks of different shapes, sizes and colours. SHRDLU has a robot arm on the screen and is able to pick up the blocks, one at a time, and move them around the world. This simple configuration represented a significant advancement over other systems. It could remember things from earlier, justify its actions and learn new things about its environment. The program was written in LISP. Some of the program's capabilities are highlighted by Winograd:

> Person: GRASP THE PYRAMID.
>
> Computer: I DON'T UNDERSTAND WHICH PYRAMID YOU MEAN.
>
> Person: FIND A BLOCK WHICH IS TALLER THAN THE ONE YOU ARE HOLDING AND PUT IT INTO THE BOX.
>
> Computer: BY "IT", I ASSUME YOU MEAN THE BLOCK WHICH IS TALLER THAN THE ONE I AM HOLDING.
>
> Computer: OK. (does it)

SHRDLU explains what it is doing before doing it so that the human user knows what SHRDLU understood from the command. When faced with ambiguity, SHRDLU can ask questions:

> Person: HOW MANY THINGS ARE ON TOP OF GREEN CUBES?
>
> Computer: I'M NOT SURE WHAT YOU MEAN BY "ON TOP OF" IN THE PHRASE "ON TOP OF GREEN CUBES". DO YOU MEAN:
>
> >   1 - DIRECTLY ON THE SURFACE
> >
> >   2 - ANWHERE ON TOP OF ?

SHRDLU is capable of experimenting and learning about its environment:

> Person: CAN A PYRAMID SUPPORT A PYRAMID?
>
> Computer: I DON'T KNOW.
>
> Person: STACK UP TWO PYRAMIDS.
>
> Computer: I CAN'T.

It can also remember what it has done in the past and answer questions about it:

> Person: HAD YOU TOUCHED ANY PYRAMID BEFORE YOU PUT THE GREEN ONE ON THE LITTLE CUBE?

Computer: YES, THE GREEN ONE.

Person: WHEN DID YOU PICK IT UP?

Computer: WHILE I WAS STACKING UP THE RED CUBE, A LARGE RED BLOCK AND A LARGE GREEN CUBE.

Person: WHY?

Computer: TO GET RID OF IT.

Person: WHY DID YOU DO THAT?

Computer: TO CLEAN OFF THE RED CUBE.

Person: WHY DID YOU CLEAR OFF THAT CUBE?

Computer: TO PUT IT ON A LARGE GREEN CUBE.

Person: WHY DID YOU DO THAT?

Computer: TO STACK UP THE RED CUBE, A LARGE RED BLOCK AND A LARGE GREEN CUBE.

Person: WHY DID YOU DO THAT?

Computer: BECAUSE YOU ASKED ME TO.

SHRDLU is grounded in its world – it has things to talk *about*. This dialogue gives some indication of SHRDLU's relations with its world, such as with blocks, physical laws, and human interlocutors. Further investigation reveals the interaction of SHRDLU with such entities as LISP, electronics, and the history and future of AI research and design.

SHRDLU interacted with many texts as well as countless academics. Winograd claimed in his dissertation that "the challenge of programming a computer to use language is really the challenge of producing intelligence."[404] But in the book *Understanding Computers and Cognition*, he and Fernando Flores argue that neither is really possible.[405] They write, citing examples of programs that attempt to produce natural language, including SHRDLU, that it is not possible for a computer to truly understand language:

> If I write something and mail it to you, you are not tempted to see the paper as exhibiting language behavior. It is a

medium through which you and I interact. If I write a complex computer program that responds to things you type, the situation is still the same – the program is still a medium through which my commitments to you are conveyed... it must be stressed that we are engaging in a particularly dangerous form of blindness if we see the computer – rather than the people who program it – as doing the understanding.[406]

This statement is a prediction that it is impossible to build a Strong AI entity and is comparable to the Chinese Room thought experiment argument. Winograd and Flores speak more of technology as something that is useful and interactive and use phenomenological language; they emphasise the tool-use of AI agents.[407] SHRDLU as an experimental actant reveals to us another way of being, the alien logic of the computer, not a logic that is comparable to but lesser than humans'.

SHRDLU is a virtual robot with no direct agency beyond the clearly-delineated walls of its system. It can be discussed as a "robot" in the same way that it discusses robot tortoises and social robots, which exist in the physical world. Some definitions of robot, such as that put forward by Maja Mataric, apply only to the latter variety, that are both "autonomous" and "exist in the physical world".[408] Karel Čapek's original usage of the word in his play *R.U.R.* (1920) described mechanical beings (*roboti*) similar to humans that work in a factory and later rise up in a rebellion, which is comparable to modern ideas of androids or cyborgs.[409] But the word itself is derived from the Czech word *robota* meaning "forced labour", and the robots in the play are actually built in a way that makes them very difficult to distinguish from humans. SHRDLU and similar entities are clearly examples of both these qualities. SHRDLU is employed in forced labour for the human user, without possessing anything that we might describe as a choice in the matter. When we look at the modern usage of the term, "robot" usually refers to an entity that exists in the physical world. A notable exception to this is the "bot", which is an entirely virtual entity which commonly performs automated tasks on the internet – a subset of this, the "chatbot", could be seen as a descendent of SHRDLU, employing natural language techniques in an attempt to appear as a human to the user. It is possible to trace a non-physical robot genealogy. We may look at the specific version of SHRDLU running on a particular computer as a position on the spectrum of robot corporeality, which spans from *Astro Boy* to Voyager 2. It can be described as a borderline fictional robot, possessing a different kind of agency than a robot that exists only within a human-driven narrative. SHRDLU possesses an extended agency, influencing elements of its personal virtual world for one thing, but also acting through the eyes and brain of the human user. We do not even need to adopt a OOO perspective to state unequivocally that SHRDLU does, in fact, exist. For OOO the virtual/embodied robot distinction is not indicative of any degree of realness in two objects. Any such hierarchy of reality is rejected in OOO. As Harman says, all objects may not be real in the same way, but "they are equally *objects*" and therefore can be discussed under the same terms.[410] They are both real, but with a different way of being in the world. It is perhaps better to compare the *physically embodied* tortoise with the *electronically embodied* SHRDLU, but to bear in mind that each is equally an object. The difference between physically and electronically embodied robots is a distraction from learning about these relations and qualities in alien phenomenology. The sensory worlds inhabited by physically and electronically embodied objects are very different, but both require investigation.

There is at least one important difference between the two types of objects, however. In a sense, both the tortoises and SHRDLU are programmed to perform exactly the same interaction under identical circumstances. If one were to feed the same sample dialogue above into SHRDLU it would respond in an identical way, both in the movements of its robot arm and its linguistic responses. Similarly, if the robot tortoise is presented with exactly the same physical situation, with identical light sources and obstacles, it would be expected to behave in the same way each time. But this is not possible in an analogue world. No amount of prior planning could result in exactly the same experimental setup each time. SHRDLU is a closed system. In thermodynamics, a closed system is that which is self-contained and does not permit matter to transfer across a boundary (although it does permit the movement of energy).[411] The conditions are set and are not influenced by external variables. A digital system like SHRDLU is constrained by its programming and will always react in the same way no matter how many times an experiment is repeated. In contrast, an open system is influenced by external variables.[412] This is true of an analogue system like a robot tortoise, which depends upon things like the ground and is affected by variations in it. For this reason, the tortoises *seem* more autonomous. When encountering a new setting the tortoise appears to make a decision about its setting based on sensory data in a way that SHRDLU is not capable of, even though we know from Winograd's commentary that the system does in fact have a very complex means of making decisions. It may be that the fact that we can in principle explain everything about SHRDLU's performance affects our conception of its semi-autonomy.

SHRDLU's electronically embodied status means that it can be reproduced in entirely different media. My own interaction with SHRDLU came through a graphical 3D version of SHRDLU with an extra Java layer remediated by an early version of Windows,[413] certainly not the original medium. The hardware on which SHRDLU was originally run is now too antiquated for the program to run on contemporary systems without alteration, even though the original LISP programming may be retained. Consequently, newcomers to SHRDLU may never have access to the original version, to the object that was called "SHRDLU" by Winograd. If the program acts in a similar way, however, there is a certain sense in which the version does not matter to SHRDLU, which exists as a concept beyond its incarnation in a specific software-hardware system. The concept might "run" on the media of many different human brains, many of which will give it different connotations, associate it with different ideas and have different opinions as to its veracity. But it is still ultimately the same concept and it exists independently of its media. The medium may be characterised as the extraneous details that make up the sensory qualities of an object – they change the way we sense the object at any given time, but they do not affect the withdrawn real object. Nevertheless, an incarnation of SHRDLU on a specific device is an object too, although this may be counter-intuitive. This is the same as the distinction made in Chapter 2 between the Kinect-artefact and the Kinect-brand, but because the programming of SHRDLU is always the same it is difficult to say where the concept ends and the specific incarnation begins. But wherever the line is, interaction with any incarnation of SHRDLU helps us to build knowledge of its sensory world, and also knowledge of a cultural and academic phenomenon called SHRDLU.

Postphenomenology is useful here as a contrast to OOO because most of the very close relations between SHRDLU and other objects in its world are confined to within the computer – and, indeed, when it was first made that computer would have been very limiting. Flipping postphenomenology around so that the nonhuman is primary is distinctly necessary in this case study since the relations are complex and interrelated but enter into postphenomenological relations with each other as well as with humans. Ihde's postphenomenology can be supplemented by alien phenomenology to result in extensive theorisation of nonhuman relations.

Postphenomenology

Comparisons between postphenomenology and ANT have been drawn by Peter-Paul Verbeek:[414] they are both non-essentialist and are grounded in a relational model. They both try to find a resolution to the insupportable metaphysical gulf between humans and nonhumans, but Verbeek thinks that postphenomenology is more "nuanced" because it goes beyond locating associations between entities and describes those associations as well.[415] Robert Rosenberger and Verbeek have called postphenomenology an "empirical philosophy".[416] This is the pragmatic reason for a postphenomenological approach with SHRDLU. The more important one here is that postphenomenology gives us particular insight into the very special, very privileged relationship between human being and technological construct. Don Ihde's postphenomenology reveals a model of human extension through technology that sublimely articulates the *feeling* of being changed by technology. Postphenomenology is an important movement in contemporary philosophy of technology and it seems necessary to present it here and critique it alongside those ideas that decentre humans.

This section will begin by briefly outlining Don Ihde's postphenomenology with some conjecture as to how the categories of experience that he describes might be used to articulate machine experience. Next, there will be a discussion of SHRDLU with relation to these categories, with particular reference to the postphenomenological notion of multistability. Finally, there will be an account of what postphenomenology can lend the notions of cultural agency that have proved problematic in previous chapters.

Don Ihde self-consciously brings phenomenology into science and technology studies.[417] His postphenomenology explores how we can consider technologies to take on the role of human intentionalities. He does not reject the ideas of previous phenomenological philosophers, but he argues that ideas must evolve in order to be relevant to other disciplines.[418] Theory has evolved over time, but perhaps more significantly technology has changed a great deal. The invention of new artefacts requires discourses of technology to enter into relations with different fields and thinkers. It may be that today the role of technology in changing our perceptions is more apparent. An example is the moral panic caused by young people overusing screen media such as online videogames. The rate of technological progress is so rapid that new generations have vastly different mediated experiences from their parents. So while the teenager playing a videogame six hours a day feels as though she is an active participant in a virtual world, society might tell her that her perceptual experiences are less real than those of a non-gamer. Yet the fantasy world inside the computer is real. We can tell, because it has an effect on the teenager and others. It also influences the computer, the internet, and aspects of human culture. It may enter into all of these relations, but in order to enter into a relation with the teenager, it must be mediated by the computer screen.

Ihde's phenomenological ideas explain how humans can have technologically mediated intentionalities. The technology becomes one of the key actors in Ihde's theory. For example, Galileo used a telescope to detect four of the moons of Jupiter, which are invisible to the naked eye. Ihde claims that this was a significant change in the way humans perceive the world, because it required people to trust in a machine to convey accurate visual information.[419] This was one of the most important moments for the shift to what Ihde calls instrumental realism: a belief that technology can augment and enhance our senses, and that those sensations are reliable. Through this transformation, what is meant for a human to sense the world changed. The object of study is not just the relationship between human and world, but the ontological trinity of human, world and technology.[420] Technology is not a passive translator of information but participates non-neutrally in the relationship. After developing these ideas over a long period of time, Ihde and colleagues eventually gave them the label of postphenomenology. Postphenomenology has become a very influential way of discussing technology because it provides tools to explain how we can experience phenomena through a machine that changes sensory input.

Ihde describes three types of relationships in which technology mediates human intentionalities. These are explored in detail in his 1990 book *Technology and the Lifeworld: From Garden to Earth*.[421] He uses mostly visual examples, so the technology mediates the human intentionality of seeing. His analysis could also be applied to other intentionalities, such as hearing (for example, with a hearing aid), remembering (with a personal organiser), perception or cognition (when one is under the influence of psychedelic drugs).[422] But it makes sense to begin with visual phenomena, following Ihde. He has elsewhere made a study of the dominance of sight over other senses in science. The increasing tendency to portray data in visual form, or "science's visualism",[423] has resulted in a privileging of sight over the other senses. Therefore, to introduce Ihde is to focus on the visual and to use examples of sight intentionalities. But machine "vision" is culturally constructed and bears limited resemblance to human vision.

Ihde identifies three types of relationship between humans and technology that change (or improve) our experience of the world. The first is the embodiment relationship. This applies to technologies that become part of the body to enhance certain intentionalities. For example, the intentionality of seeing is enhanced/changed by wearing eyeglasses. The body's capabilities are transformed by the artefact. The object becomes a part of the way the human sees. This process is not limited to simple machines. Hearing aids exist in the same kind of relationship as eyeglasses but are much more complex.[424] Maurice Merleau-Ponty's blind man's cane is an embodiment relationship, as it is a direct means of extending the tactility of the body.[425] Ihde depicts this relationship as:

(human-technology) → environment

The technology is taken into the human to permit the experience of the world. The object is hardly noticeable, and in fact performs better the less obtrusive it is. If a technology is good, it "withdraws".[426] It becomes "quasitransparent".[427] (This withdrawal is similar to the withdrawal in OOO, or to Latour's "black boxes"[428], in that the means of accessing the technology are removed from external actants, although Ihde's use of the term occurs instant by instant whereas Harman's withdrawal is eternal. The terms are not to be confused in origin or in application.) Ihde poignantly describes withdrawal as a fundamental contradiction inherent to humans' use of machines. We want the power that comes from the technological transformation, but we secretly desire "to escape the limitations of the material technology."[429] Thus, embodiment relationships reveal the human-technology hybrid to be a site of ambivalence, demonstrating the human craving for direct experience.

This craving is revealed to us through the narrative of verisimilitude in the history of visual images.[430] Many art historians have identified a persistent tendency to produce more and more realistic visual media throughout history. This culminates in the ideal of virtual reality (VR), a concept that remains elusive to us and will remain elusive for the foreseeable future. Ken Hillis says that a true VR mediation would "pass for or merge seamlessly with perception itself",[431] taking the place of not only visual stimuli but sound, smell, fatigue, hunger and all other affects. It would not even be possible to detect the pressure of the headset on one's skin. This ideal remains infeasible for VR researchers in the short term. In Ihde's terms, we seek technologies that will transform our intentionalities, but we wish to erase all traces of the mediator and become utterly embodied. Yet there are also alternative narratives of media. Art historian Jonathan Crary points to places in the history of images where humans have craved mediation through images.[432] Impressionists and post-impressionists, for example, deliberately distorted visual stimuli, leaving the greater truth behind their work to be interpreted by the viewer. These are pieces that disdain the natural human desire for embodied mediation and require a hermeneutic process (see below).

A comparison to the human-centric embodiment relationship for considering machine phenomenology is necessary at this point. This relationship must feature a technology making use of a human actant to perceive the world. In the case of a desktop computer, an embodiment relationship is formed between typist and keyboard. To rephrase Ihde's model for this perspective:

(technology-human) → world

The keyboard is, along with the mouse and the network, one of the most important ways that a desktop computer experiences the world. The depression of certain keys acts as stimuli to computer receptors. Each receptor has, depending on context, a different effect, rather as acetylcholine acts only upon acetylcholine receptors, yet it may have multifarious effects on different tissues. For example, sometimes my laptop asks me to enter a password to access the university Wi-Fi, since it has detected that it cannot connect to the internet and perform its tasks without it, and since it has also failed to remember the password. It persuades my brain and fingers to work by telling me that it is having a problem, in language that I can understand. It thereby extends its own capabilities using my body. The computer needs the human user to depress keys in order for it to function, and makes use of human hands (and a human brain) to fulfil its purpose. Other objects involved include wires, key codes, and a keyboard driver.

Using the word in the sense of "reading" or "interpretation", Ihde next discusses hermeneutic relationships. This category refers to objects that transmit data or information which must then be interpreted by the human user. The most basic example in Ihde's account is of a human reading a chart. The chart is inscribed by a human, but alone it is useless. It requires a human with the knowledge and understanding to read and interpret it. It refers to events that occur beyond its material self.[433] Another example is the thermometer that detects the temperature of the air so that the human reader can hermeneutically receive sensory information. The machine sensation is transcribed into a written or visual form which is then reconstructed by the human. This is depicted as:

human → (technology-world)

The technology is experiencing the conditions of the world. It has a process by which it acquires information, such as through sensors or human input. But the human user interpreting the technology has no direct access to the information that the machine reveals. The human must have an implicit understanding of what is indicated by the machine. In the thermometer example, it is no use to simply have a machine that tells us the temperature is thirty-five degrees centigrade. It requires a human user to understand what the value "thirty-five" means. On a weather report, it indicates an uncomfortably hot day, while on an oral thermometer it is a dangerously low temperature for a human body. And as Ihde points out, a native of North America would also need to know how to convert centigrade temperatures into Fahrenheit values before they could be read properly.[434] As in embodiment relationships, we can extend this idea to more complex examples. The Parkes Radio Telescope measures hydrogen gas densities in other galaxies.[435] The phenomenon under investigation is both distant and invisible to human senses, but the telescope is capable of displaying the data in visual forms, with different colours to indicate densities. The machines translate data into a form that may be understood by human beings, so long as they possess the knowledge to interpret the images. Note that it is explicitly implied here that nonhuman sensation not only exists, but that it is possible for humans to gain information from it. In other words, the human can understand what occurs in the telescope's sensory world, to a limited extent.

Examples of computers interpreting human actions abound. The writing of code or algorithms to control processes results in human texts that the computer must interpret and act upon. The computer depends upon the accuracy of the human typist to reduce errors in the human-machine process of semiosis. While writing code, the omission of a semi-colon may result in a non-functioning program. The computer relies on the human to produce good, comprehensible code. After being converted into machine language, the instructions become a facilitator of machinic action:

> technology → (human-world)

Poorly-coded programs rely overmuch on human intervention or make it difficult to make future alterations. Well-coded programs are so efficient and effective that they are blackboxed and become invisible. When SHRDLU performs as hoped, its programming is invisible to all but the most specialised user because no errors in comprehension or processing are evident to betray the skilful tricks that Winograd may have used to program the system. A corollary exists in the practice of providing automated voice-controlled telephone services. A simple, well-organised service relies on responses of only "yes" or "no" and allows the caller to complete their business swiftly. Yet all too often the service requires users to provide complicated answers that are difficult to decipher using a voice-to-text program. In these cases, the caller is typically (or at least hopefully) referred to the human operator. The goal in any kind of automated system is to remove the need for human intervention. For example, the goal of a personalised Google algorithm is to remove the need for the user to refine search topics. Google learns about the user's location and other information to provide more relevant local search results so that the user does not need to sift through too much material. This also allows Google to target advertisements based on what it knows about the user. The ability to interpret is hindered by human errors, such as entering one's location as "Austria" instead of "Australia".

Alterity relationships occur between humans and technology that we consider to be an "other" to which one can relate. The term comes from Lévinas's notion that one person is radically different from any other.[436] In Ihde, the "otherness" is extended to semi-autonomous machines like certain robots, although he frequently qualifies it as "quasi-otherness".[437] The relationship is depicted as:

human → technology-(-world)

The parentheses indicate that although the relation is primarily between the human and the technology, there may also be relations through the technology to the world. One of the difficulties for Ihde in describing this relationship is that the concept of technology as an "other" varies. What it means to anthropomorphise a machine is not uniform across cultures.

In theory, an alterity relationship should be bidirectional even in a non-OOO sense. It is a relationship that acknowledges the otherness of technology. A true alterity relationship would be the same from both sides of the interface. These kinds of relationships can also apply to computer-computer relationships that occur within a network as well as human-human relationships. Yet if we include these as alterity relationships, then surely we must also account for human-human interaction. Perhaps the best way to define the interaction is using an unqualified account of object relations, regardless of anthropomorphic status:

object A-(-world) → object B
object B-(-world) → object A

The relationship is bidirectional. But it is also asymmetrical. Object A's ability to impart knowledge of the world to Object B might be different from Object B's ability to impart knowledge to Object A. For example, a desktop computer and a smartphone may be in an alterity relationship, connected by a cable. But each can only impart certain types of sensory experiences to the other.

Embodiment, hermeneutic and alterity relations are the three major means by which technology mediates human intentionalities in Ihde's construct. Added to this are also background relations, which are relations between humans and automatic technological processes, such as climate control. The background relation may only briefly be directly set up by the human, for example the setting of a thermostat. For most of the time, the technology operates as a "present absence"[438], affecting the human's environment but without direct contact of "focal"[439] technologies.

Ihde's categories represent a significant departure from critical theories in which objects are open for interpretation, particularly with alterity relationships. He flattens hierarchies of objects: asking a question of a robot designed to tell where to find things in a store is as worthy of attention as objects capable of a multiplicity of tasks such as a personal computer. Doubtless the relationship between computer and human user is determined by the particular application in use at any one moment. Some applications are hermeneutic; for example, a web service that tells the user about the temperature in another city. This application requires human interpretation to become useful. But in other states other kinds of relationships may be formed with the computer.

Interpretations of machines by humans are sometimes beyond the engineer's intentions. A computer or a robot is a cultural artefact as well as a semi-autonomous being. It is, as Ihde frequently points out, inextricably linked to the culture that produced it. It is sometimes not as easily interpreted as a painting or a film, but it is still a product of our cultural imaginary and can be a mediator of ideas, norms, taboos, practices and identities. For example, a social robot may be designed to be relatively independent and to interact freely with human beings in, as Ihde would say, an alterity relationship. But social robots represent certain cultural ideas. Chapter 5 specifically notes a tendency for "feminine"-looking robots to be objectified by their creators and users. When we talk to a "feminine" social robot, an "other", we are also receiving coded messages about social interaction and our relationship with robots, and also about relationships between women and society. Is this, then, simply a very complicated hermeneutic relationship? It is at this point that Ihde's categories become entangled.

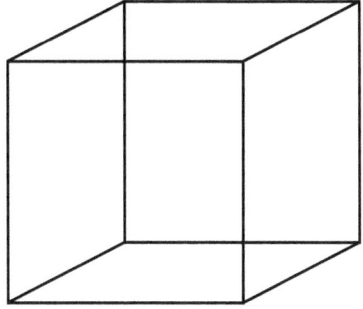

Figure 2: Necker Cube

Is it possible for a complex machine like a computer to form an embodiment relationship with a human? If a Youtube user can access a video taken with a smartphone of protests in Cairo, is the computer then extending their vision? The mediating qualities of the screen are a problem because the video is never as reliable and firmly embodied as the eyeglasses – after all, the footage is only ever a brief snippet and is a poor replacement for being present at the scene. But Ihde is aware of the significance of the frame around mediated experiences. In describing Galileo's observation through a telescope, he points out the ontological leap that must be taken to accept the screened space as real. Framing an image immediately makes it seem less real to us (or real in a different way). We are aware that a video on CNN may not be showing the whole story. Perhaps then the user is in a hermeneutic relationship with the footage. It requires interpretation, context and prior knowledge to combat the disembodiment, just as a thermometer needs a numerical scale to allow us to interpret the expansion of mercury and knowledge of, for example, the temperature that indicates a fever in the body. The question is an old one in the study of media: is it better to allow the viewer to become immersed and to pretend that there is no mediation, or to provide lots of interpretation as compensation? We can see by this that Ihde's categories rely heavily on how we socially construct embodiment.

Here the concept of multistability is useful, a means of acknowledging that relations between humans and machines vary by the instant and by the person. Ihde's approach to this idea begins with an image such as a Necker cube – a two-dimensional interconnection of black lines which can appear to the viewer as one of either two cubes, but never both at the same time (and, as Ihde says, can be seen as other shapes as well).[440]

This is a metaphor for the relations between humans and technology. A technological artefact may be used in different ways by different people, in different contexts, and will mean different things to different people, often beyond the intentions of their creator. Robert Rosenberger describes the ways that a public bench becomes not simply a place to sit, but also a bed for the homeless: "[i]n this way, a bench is a multistable technology with both bench-as-seat and bench-as-bed stabilities."[441] Evan Selinger, however, criticises Ihde's heightened ambivalence about artefacts and context, pointing out that he has sometimes been accused of an overly apolitical approach to technology.[442] Multistability can be a dangerous concept when it is applied to objects that pose significant risk to human life. Furthermore, sometimes multistability may be desirable but for practical reasons it may not actually occur, perhaps because there is not enough time for humans to find novel applications for the artefact. Nevertheless, the concept of multistability prevents an overly deterministic view of technology and allows for the demonstrable discovery of novel ways for humans to use nonhuman objects.

Shifting categories of human-technology relations are particularly evident in screen media, such as in the first-person shooter game *Halo 4*.[443] Games particularly complicate these categories as the player must engage in all of them and sometimes all at the same time. Because videogames are not completely immersive, the human player cannot be totally embodied and take on the role of the digital Master Chief protagonist. Nevertheless, the human does have some embodied experiences. When the player moves Master Chief's head, the view of the virtual world shifts as though we are looking through his eyes. When struck a heavy blow, the console's controller shakes and rumbles, to give us a sense of what Master Chief is experiencing.

These simple innovations are designed to extend human sensory systems into virtual spaces.

In terms of ANT, quasitransparency is only achieved when both human and virtual actants are cooperating. The game is successful because the player's goal and the character's goal match and because each is capable of affecting the other's agency in a predictable way. This is particularly evident in a game like *Halo 4* because the virtual actant is an anthropomorphic avatar. The player views the world from behind Master Chief's eyes. Consequently, the player's action is immediately transferred into Master Chief's "muscles" and Master Chief's "eyes" transmit information immediately to the player's eyes. It is therefore essential that player and character have allied goals, and this is typically built into the narrative form of videogames as well as being hard-wired into the code. Master Chief has a goal: for example, he must fight his way through hostile terrain to rescue his comrades. Fortunately, the player's goal is aligned with the character's goal through the embodied mediation of the avatar. The typical player wants to move Master Chief in a skilful way to overcome obstacles, such as enemies or puzzles, and thereby progress in the game. The impetus behind the decision to move through the jungle is different. Master Chief wants to perform a rescue. The player (probably) wants to beat the level and advance the story. In other words, each translates the other in order to bring their goals into alignment. Because player and avatar are allied, neither resists the other's decisions and the result is a satisfying feeling of embodiment. Because of this alliance, the game does not require Master Chief to disobey the orders conveyed to him by the player, and he performs actions corresponding to the player's gestures (so long as other actants, such as the controller battery, are also enrolled, to put it in ANT terms). The player, in turn, uses gestures to try to aid Master Chief in his mission. The relationship between player, avatar and character is not always as straightforward as this example and varies in degree.[444] However, the logical alliance provides a victory/loss model that is resisted in relatively few videogames. A notable exception is the common tactic in activist videogames to complicate the implied cooperation between player and avatar. For example, in activist videogame developers Molleindustria's online game *Operation: Pedopriest* the "aim" is to help rapists escape justice.[445] Clearly, the real aim of the game is to foster awareness of something that the developer considers to be an important social issue. (Bogost calls this "procedural rhetoric."[446]) In this example, the player's embodiment is limited because the avatar fails to integrate them into its network. The sense of identification with the avatar is limited at best. This is not a problem for Molleindustria, but a AAA Xbox 360 title must aim to cooperate as much as possible with the user in order to achieve a sense of immersion, and therefore for the game and the Xbox 360 to succeed in their goals of being bought and enjoyed. The *Halo* series has been very successful, and can only be played on Xbox consoles, so Master Chief's successful formation of an embodiment relationship with various players indirectly benefits the Xbox and Microsoft brands.

Intuitive hermeneutic and alterity relationships are also vital to the success of the game. As an example, we cannot feel pain as Master Chief loses ground in a fight, so *Halo 4*'s interface provides us with a health bar. Master Chief cannot continue on his mission if he dies, so it is important for the game (which is strongly allied with Master Chief) to convey the state of his health to the player through a hermeneutic relationship. (The actual experience of Master Chief's death could be considered an embodiment relationship.) The player must read and interpret the health bar, and the other diegetic and non-diegetic devices on the interface, and the game must provide accurate data. Furthermore, we could say that the player has an alterity relationship with Master Chief. He is undoubtedly an entity different from the player who essentially exists to guide us through a virtual space and story. His lifeworld is not the player's lifeworld. Moreover, Master Chief enjoys brief periods of autonomy in cut-scenes during which the player has no control over him. Ihde's categories are open to far more interpretation than they at first appear, and individual relationships may both move between them and transcend them.

The human experiences the game, and the game experiences the human through the input of the controller. The depression of controller buttons corresponds to certain pre-programmed responses in the videogame console. For example, shifting one of the joysticks to the left causes Master Chief's head to rotate, allowing the player (and Master Chief) to view a different part of the virtual environment. In other words, the machine senses the mechanical change in the controller which is then sent wirelessly to the console. Here, the sensation is "perceived" and the console interprets the movement. The activation of code that changes the view of the virtual environment is only enacted after this interpretation takes place. This analysis allows a glimpse into the curious digital world of the console and gives an insight into the agency of the machine. We can also see how the interaction is bidirectional and asymmetrical. Master Chief's sensation of the player occurs concurrently with the player's sensation of Master Chief, but each occurs through different means and presumably results in very different perceptual experiences. SHRDLU too possesses several means of mediating human intentionality, and it occurs in embodiment, hermeneutic and alterity relations.

SHRDLU and Ihde's categories

It is important to bear in mind here that Ihde's categories are ways of accounting for the extension of human sensation or perception by technology. Similarly, when we apply these categories to relations between SHRDLU and other nonhumans it is worth noticing how SHRDLU extends or transforms aspects of those nonhumans, and vice versa. This means making assumptions about the intention and agency of the human user, so there has been some attempt here to represent various possible uses to which SHRDLU might be put, in accordance with the notion of multistability. For example, the purpose of SHRDLU (from a human perspective) might be to allow the user to have a very detailed, albeit dull, conversation with a computer in a "natural" way. Another purpose is to test the hypothesis that NLP is only possible if the AI is grounded in the world. With an emphasis on the visual aspect of SHRDLU's presentation, it might be inferred that the user employs SHRDLU to gain a representation of another world, to see what SHRDLU sees. SHRDLU could also be employed as a voice-controlled means of moving articles around in a warehouse, if it were allowed to form relations with the right kinds of equipment. Yet another perspective reveals SHRDLU to be a means of progressing Winograd's career, and later to provide Dreyfus with an example of AI limitations. To see how SHRDLU extends and modifies these human programs of action, it is necessary to first delve further into the world of SHRDLU.

The first thing we must consider in building a picture of SHRDLU's environment and relations therewith is the wider external world. This is the best way we have of learning about its qualities and relations, and therefore of asking SHRDLU about its sensory world. SHRDLU was built at MIT initially as a part of Winograd's doctoral research. The conditions under which Winograd was allowed the equipment and time to construct the system are material to its existence. Similarly, it is important to take a broad view of AI when considering this question. Perhaps SHRDLU would never have existed had there been no optimism around computer language use born of things like ELIZA and translation programs, or no trend toward the development of microworlds in AI. And of course, SHRDLU's existence in turn has had a significant impact on subsequent AI projects. All of these objects that are external to the computer world in which SHRDLU exists – the lab, the university, Winograd himself and the environment of AI at the time – are objects with which SHRDLU has significant relations, and cannot be discounted when we ask what it is like to be SHRDLU.

Of direct interest though is the internal state of the system. SHRDLU has a high understanding of everything in the block world but at the cost of that world being tightly constrained. The closed nature of the system ensures that the microworld is tractable and always performs the same actions in the same way, and the constrained nature of the system permits all experiments to be predictable. To ask, then, what it is like to be SHRDLU is to have a slightly better chance of success than to ask what it is like to be Grey Walter's tortoises, since all factors are contained and external variables are minimised. Yet the internal consistency evident to a human interlocutor conceals SHRDLU's way of making decisions, which is quite complex. The representation on the screen conceals the vast number of objects under the surface. SHRDLU is an assemblage of processes which, combined, achieve a goal, as in Marvin Minsky's portrayal of a system.[447] We can set out to create a "builder" program – a program that aims to build a tower out of blocks.[448] But the program requires sub-programs, which then require other sub-programs beneath it. Every time SHRDLU acts in response to a question it engages with a vast structure of sub-programs that lead it to make a decision about what it should say and do.

We can interact with SHRDLU in a way that is relatively simple for humans, but only because of SHRDLU's immensely complex programming. SHRDLU's understanding of language is described by Winograd as being constituted by three layers:

- A syntactic parser which works with a large-scale grammar of English;
- Semantic routines which embody the knowledge needed to interpret the meaning of words and structures; and
- Cognitive deductive system for exploring the consequence of facts.[449]

These systems, in addition to the block world and the set of programs for generating English responses, constitute SHRDLU's world, and at a deeper level it is also constituted by LISP, logic, and the fundamentals of computing. These are all real qualities that define SHRDLU's way of being in the world.

Any action by the system is the net effect of a vast array of interwoven actants. In a 1972 paper, Winograd depicts the various interrelating parts of SHRDLU in graphical form, showing how elements like "Dictionary",

```
                    Pressure; elevation
            ──────────────────────────────►
Robot arm   ◄──────────────────────────────   Block
                    Hardness; weight
```

"Input", "Monitor", ""Programmar", "Semantics", "Grammar", "Answer", and "Mover" link to one another.[450] PROGRAMMAR is the language created for expressing the details of grammar in the system, and PLANNER is a language for problem-solving procedures, and MONITOR is a program that calls the basic parts of the system.[451] The other parts of the system are named for the function that they perform. Winograd's 1972 paper on

```
        Imperative to move in visual representation; virtual movement
            ──────────────────────────────►
Robot arm   ◄──────────────────────────────   Block
            Virtual presence; capacity to carry out instruction
```

SHRDLU goes into detail about each of the components, and explains which performs tasks that may not be obviously accounted for. For example, the ANSWER component is responsible for remembering discourse as well as for creating responses. It can be seen from this that the internal world of SHRDLU is one constituted from many disparate parts and that the seamless unified responses that it offers to questions and statements are something of an illusion. As Douglas Hofstadter writes, "[t]he program is like a very tangled knot which resists untangling; but the fact that you cannot untangle it does not mean that you cannot understand it."[452] SHRDLU is made up of many different parts, each of which is dependent upon layers of sub-programs.

But we must remember that the presence of many different parts does not reduce the wholeness of SHRDLU as an object, or the wholeness of the ANSWER component, or the wholeness of the robot arm. So what is it like to be the robotic arm in SHRDLU? This question has the advantage of offering conceptual symmetry with the robot tortoises and other objects of study in this book. It is relatively easy to consider the robot arm's relations with other objects without asking where the object of study begins and ends, which is not necessarily true of the object or assemblage of objects signified by the word "SHRDLU". When we ask what relations exist for the robot arm we can identify a pleasingly diverse list of objects: the blue block, the block-world table, the human user and their input, the syntactic parser, semantic routines and cognitive deductive system, LISP, previous block movements, the computer and the electrical current, *et cetera, et cetera*.

The robot arm's most obvious contact is with the blocks and world of blocks. But it is of a different nature than the contact between a real robot arm and block world. In an interaction of a physically embodied robot arm and block we might expect to see a bidirectional relationship of this sort: The robot arm imposes pressure on the block and causes it to rise off the surface on which it was resting. Simultaneously, the robot arm senses the hardness and weight of the block.

Instead the bidirectional relationship in SHRDLU looks a little more like this:
The robot arm compels the block to move (virtually) to a new location. Simultaneously, the block is tractable to virtual movement by the robot arm.

Doubtless there are many other factors that are not immediately apparent to the human user. But in each case, the bidirectional relation is asymmetrical. Each object gains only a caricature of the other object, as Harman would put it. We cannot be certain of the nature of this caricature, although Winograd's technical reports may offer clues.

Thus the sensory equipment given to the robot arm is primed for interaction with the blocks. Through the actions and interactions of its code it senses the fact that it is holding a block and it senses the fact that it puts it down (we know this because it informs the system, which remembers it and can tell the human user about it later). Notice that this interaction is quite different from the sensual object that would be formed between a physically embodied robot arm and block, although SHRDLU attempts to model this interaction with realistic physics. This interaction can be viewed through the visual representation that sits alongside the teletext interaction, but the true robot arm exists within the system, sensing blocks in a thoroughly non-visual way. Apart from the block interaction the robot arm is sensitive to instructions from the rest of the system which are related in machine language. These are symbolic interactions that it is hard for us to imagine. What is it like for part of a system to sense symbolic instructions from another part of the system? This is one of those thoroughly alien interactions that draw the mind away from the contrived anthropomorphism of SHRDLU as a whole and emphasise the impossibility of ever truly breaching the inner nature of a nonhuman object.

The robot arm is inextricably linked with each of the system components. It is part of a large assemblage of objects united in the goal of *being* SHRDLU, as in Marvin Minsky's explanation of assemblages of components needed to create a virtual block tower builder. Our sensual impression of SHRDLU is far removed from its true internal state, which is whole despite consisting of interactions between the objects in the assemblage. All of our information about the world of SHRDLU is conveyed through the graphical display and the text that appears on the screen. These give the impression of an agent that does not exist, a unified "player" agency that in many ways resembles a human player using the blocks and responding to questions. And this is, of course, Winograd's intention. SHRDLU the "builder" to which we relate is a sensual object, a human thing, our caricature of an alien world (or at least, the caricature available to those of us who are not computer scientists engaged in working on SHRDLU).

Another interesting relationship relevant to this discussion is that which exists between SHRDLU and the English language. Surely if SHRDLU is in a relationship with a language, it is an embodiment relation. Words are firmly embedded in SHRDLU, and it actively manipulates language. We can say with some confidence that the English language is an object. It is an entity with identifiable qualities that distinguish it from other entities. It clearly enters into relations with other objects such as texts, thoughts and utterances, and is affected by them as well. It changes over time but remains recognisably the English language. And it is in a bidirectional relationship with language-performing objects like SHRDLU as well. SHRDLU is clearly a user of the English language. It is shaped by qualities of the English language such as its syntax, signifiers and alphabet which are essential elements of its functioning and purpose. SHRDLU takes on these qualities from its association with the English language, as though one were physically moulding the other. In OOO there is always a reciprocal (but different) interaction. What effect could SHRDLU, or any language-performing object, have on the English language? Well, firstly SHRDLU happens to represent a significant innovation into the study of language, still shaping the way that we think about how computers can use language and even how language works in humans. But secondly its role as a language user is similar to that of a human language user in that by communicating by using particular words and phrasing it has a tiny but not insignificant effect on the English language as a whole. Regardless of whether SHRDLU is a true user of language, SHRDLU as a text is in a bidirectional relationship with the English language, as when a note contains a word that is unfamiliar to the reader, causing the reader to integrate the word into their vocabulary and give that word a minutely greater prominence within the English language. SHRDLU has a relationship with the English language in the same way that Walter's texts on tortoises affect the field of cybernetics and robotics, not just as media through which Walter speaks.

Given this overview of SHRDLU's world, it is easier to see how SHRDLU forms the types of relations that Ihde and others describe. Its embodiment, hermeneutic and alterity relations will be described here. The one kind of relation not discussed here is background relations. Background relations are an addendum to Ihde's original three categories, and they are not a great factor in SHRDLU's understanding of the world. That is, perhaps SHRDLU is often in a background relation with humans, but it is always in a non-background relation with *something*. If one walks away from the computer leaving SHRDLU running, it is still interacting with the rest of the computer. It might be possible to argue that SHRDLU's influence on the development of AI research constitutes a background relation, but this chapter discusses that issue in different terms.

Alterity and SHRDLU

The nature of alterity in SHRDLU's relations with humans is highly dependent on the writer, as is the case in all examinations of AI (one might even say that all writing on alterity is highly dependent on the writer). Clearly, the goals of the SHRDLU project intended an alterity relation between SHRDLU and a human. The computer refers to itself as "I", and it refers to the user as "YOU". The goal of reproducing "natural" speech in the form of a conversation inevitably leads to an alterity-like construction. Mark Marino refers to this process in ANT terms as "punctualization" (the blackboxing of large parts of networks and the creation of larger actor-networks),[453] commenting that it is common to assign the rudiments of an identity, including a gender, at the text or disembodied voice stage. Marino further argues that once a body is given to the chatbot, its ethnic or racial identity becomes apparent (see Chapter 5).[454] Of course, this is mostly avoided with SHRDLU, since its "body" consists only of a virtual robotic arm, which is a major alienating factor when interacting with it. SHRDLU could easily have been a child, with a gender identity and a human hand. But this would have detracted from the pure proposition-testing nature of the experiment and would have represented a similar attempt at trickery as that implicit in the DOCTOR program. Nevertheless, SHRDLU relations have a hint of alterity about them. As long as the user talks to SHRDLU about the block world, it is capable of maintaining an impression of otherness. More recent chatbots, which produce a similar language-using effect to the human user, are much more advanced and are indeed capable of luring unwary internet users into giving away personal information. In principle, the natural language conversation model is perfectly capable of producing an alterity relation.

But the fact remains that when one speaks to SHRDLU, one knows that one is speaking to a computer because of its blinkered obsession with blocks. Consequently, the question of alterity is a matter of personal opinion, and is multistable in any case. In fact, there is some evidence that alterity is a product of cultural conditioning. Cathrine Hasse has investigated this issue with regard to Paro the baby seal robot developed for patients suffering from dementia.[455] She shows that personhood is not uniform across cultures, and therefore that relations between humans and other or quasi-other entities are culturally contingent as well. In a postphenomenological sense, Paro is multistable: it is used in different ways by different people at different times. Hasse suggests a conception of technology as being a generator of signs, and the same object creates different signs for different people.[456] Hence, for an AI researcher from the early 1970s, SHRDLU might be seen to represent a brand-new type of intelligence, whose grounding in a world gives it a genuine otherness. A modern AI researcher, who is familiar with more advanced chatbots or robots might look at SHRDLU's limited block-world and fail to find any sense of otherness in its conversational skills. A child might think of SHRDLU as another intelligence, only to grow up and be less impressed by it. A person with an animistic upbringing might be in more of an alterity relationship with SHRDLU than a person without an animistic view of the world. All we can say for certain is that SHRDLU aimed to point humanity in the direction of apparent alterity. Where this alterity fails to appear, we might find ourselves in more of a hermeneutic relation. Rosenberger relates failure in computers to create connections to Heidegger's broken hammer example, citing the way in which a user might forget that she or he is using a computer until something goes wrong, like a website taking too long to load.[457] This causes an abrupt shift in transparency. Of course, the sort of malfunction Rosenberger is thinking of could be considered desirable events in SHRDLU, since many users would be hoping to detect problems or inconsistencies. In this context, a "malfunction" would be more evident in the ELIZA program that seeks to erase its own machinic identity.

To turn around Ihde's categories is to begin a speculative work; making informed guesses about what SHRDLU is likely to experience, as in alien phenomenology. We might assume that SHRDLU has a sense of alterity in its interaction with us, but it is probably not a special relationship. SHRDLU's place in the world is (or at least was) contingent on the work of engineers like Winograd and is multistable. But in its interaction with most people, SHRDLU is interacting with an entity other than itself. Its minuscule knowledge of the outside world reinforces this. Its only impression of a human user comes from the few words typed into its textual input. Of course, SHRDLU's world is so tiny that this textual input is large by comparison, but it has no other means of knowing the user and no way of interpreting the user as a co-constructor of its existence. This means that any user of SHRDLU's textual input would have an alterity relationship with it. If the output of the DOCTOR program was related to SHRDLU, SHRDLU would have limited success in making sense of the text but it would still constitute an alterity relationship. A cat, falling asleep on the keyboard, would be in an alterity relationship with it. A random number generator would be in an alterity relationship with it.

Since this is an exercise in extending the concept of alterity to nonhumans, it would make sense to ask whether SHRDLU was in an alterity relationship with the nonhumans to which we humans are accustomed. If a computer has any conception of an "other", it would not logically be limited to human others. But it would not be in an alterity relationship with shrubs, kiwifruits or statuettes in the same way that humans do not usually form relationships with radio waves unless it is through another device. The object must appear in SHRDLU's environment in order for it to form an alterity relationship. SHRDLU must be open to relations that are prone to taking the shape of alterity relations; its susceptibility to alterity relations is a consequence of the real qualities that exist within its withdrawn self.

Hermeneutic relationships with SHRDLU

If a person is unable to attain a sense of alterity with SHRDLU, the relation at issue is likely to be a hermeneutic one. A person might, for example, feel no sense of otherness about SHRDLU but be able to see the program as a way of gaining insight into the code, the system, the programming language, the programmer, the English language or the concept of artificial intelligence. Like the alterity relations, hermeneutic relations are multistable, so different people would experience different kinds of hermeneutic relations, and at different times and places SHRDLU would experience different kinds of hermeneutic relations as well. The creator of the system, or somebody else very familiar with its technical details, would be able to use SHRDLU to detect problems or limitations of the technical frame. A linguist would have a different interpretation, and so on. Here the implications are very great, because the types of interpretation to be made from SHRDLU are immense for humankind. They concern our understanding of things like language, mind and phenomenology.

The most basic hermeneutic relation lies in reading and understanding what SHRDLU writes; it means comprehending what it means by "BLOCK" and "PUT". Beyond that, human interpretations of SHRDLU are generally about using the limited available information, textual and visual, to gain insights into SHRDLU's capabilities and qualities. This is what it was built for, after all. A textual analysis will reveal, to a greater or lesser extent, what SHRDLU can do. A user can use it to test various premises. But we are totally reliant on what SHRDLU reveals to us in its text and graphical representation. The representation of the block world is not tied to any real-world block world. It is an abstraction designed to represent events in SHRDLU's system that are not readily accessible to us in any other form, and it is only connected to SHRDLU's code for our convenience. The block world image allows us to see an entirely different kind of information, rather as the radio telescope converts radiation into numbers, and then into coloured images.

The kinds of hermeneutic relations formed by SHRDLU are more limited. As described in the Chinese Room experiment, it is controversial to claim that technological artefacts are capable of semiotic exchanges with the environment. However, we can draw comparisons between SHRDLU and humans that can encourage us to interpret SHRDLU's relations as hermeneutic. SHRDLU has a hermeneutic relationship with the text added by the human. Each part of the text must be relayed to those parts of the program designed to parse the English instructions or questions given to it. It is also good at forming hermeneutic relations with the component parts of its own program. Each part of SHRDLU must communicate with other parts of the program, but only in terms that other parts of the program can understand. Relaying the instruction to "move the red block" directly to the image of the block world would result only in bafflement. The instruction must first be translated by different parts of the program and relayed to the block world in machine language. A similar process occurs when a question is asked of the program, requiring it to call its own memory and explain why it acted the way it did. SHRDLU is not as open to forming hermeneutic relations with the world as it is seen by humans, so we can again only make educated guesses.

Embodiment and SHRDLU

The point of SHRDLU for human users, unlike many computer interfaces, is not to become invisible. Very few users would play with SHRDLU for the simple pleasure of moving blocks around the screen. SHRDLU's interface exists to demonstrate the benefits and limitations of its system. In that respect, we see through the interface into the inner workings of the computer. To the serious researcher, SHRDLU's textual and graphical representations could be transparent conduits to an understanding of the program behind them. In Rosenberger's account he integrates the concept of embodiment with Heidegger's tool use analysis. He describes the way that the computer itself – the keys, screen and mouse – all fade away when one is engrossed in work, and how the computer reasserts itself when it malfunctions.[458] This is similar to Harman's analysis discussed in Chapter 1. The difference, of course, is that in OOO other objects also experience the computer as ready-to-hand and present-at-hand.

SHRDLU extends our vision and imagination beyond the human world into the world of the computer. However, SHRDLU lacks the qualities typical of Ihde's examples of embodiment relations. Unlike the glasses or the dental probe, SHRDLU is not really extending the human user's existing senses, such as vision and touch. As discussed above, SHRDLU's nonhuman materiality never fades from view (that is not its intention). There is a sense, however, in which SHRDLU allows us to extend our human senses, with reference to certain transhumanist discourses. There are aspects of the experience of SHRDLU that could be considered embodiment relations. The embodiment relationship between skilled typist and typewriter was identified by Merleau-Ponty in 1962.[459] This, as Brey points out, is not a case of the typewriter extending the typist's perception but is another type of embodiment relation that does not clearly fit into Ihde's use of the term.[460]

Embodiment relations require a firm conception of a body and experience, which is difficult to define in SHRDLU, existing as it does as code, as a way of organising a computer, and as an idea. Yet we may identify an embodiment relationship as existing between SHRDLU and the keyboard through which we communicate with it. The keyboard is a way of extending the sensory capabilities of SHRDLU, just as the telescope extended Galileo's sense of sight. SHRDLU is primed and ready to receive questions and instructions in English. But the English-speaking human is invisible to it, and only made visible through the depression of keys. It relies on the computer's software to convert keystrokes into language, but it does not perform this interpretation itself. It can be argued that the conversion of keystrokes into language is a part of its sensory pathway. Parts of our brain perform similar work when it encounters noise, converting it into language before the words themselves become clear to us. The noise-ear-nerve-brain pathway relies on interpretation (or translation, in ANT terms) at the points between objects. Similarly, SHRDLU's sensation of the world involves different moments of interpretation (or translation), but that interpretation is a natural part of SHRDLU's way of being in the world. This complicates an earlier assertion – how can SHRDLU be in an embodiment relationship with the keyboard if it is in a hermeneutic relationship with the text? Making statements about the kinds of relations SHRDLU forms implies a far greater knowledge of its way of being in the world than we can rightly claim. These words can only be metaphors and guides. SHRDLU lacks an identifiable body, so the metaphor of embodiment simply hints at an alien and disembodied process of sensory extension.

Consequences of and problems with reversing Ihde's relation types in SHRDLU

It is not easy to force these categories onto SHRDLU, and it would be even more difficult to apply them to another kind of object, such as a rock. What kind of hermeneutic relations does a public bench have, and isn't it presumptuous to try to guess? This is not a problem for postphenomenology, which is explicitly a field that discusses the effect of technology on humans, consistent with its origins in phenomenology. But it may be a problem for the nonhuman turn. It is a problem with only a partial solution; the exploration of nonhuman ways of being in the world, the application of alien phenomenology to gain a biased, human-centric idea of the real qualities inherent in the nonhuman object. Alien phenomenology allows us to speak in broad, generalised terms about the sensation and experience of nonhumans. Meanwhile, postphenomenology allows us to make sense of those sensations and experience.

Merely to say that an object has a relationship with another object is a fairly empty statement. We would immediately want to know what kind of relationship, and leaving out the capacity to *describe* relations is a serious limitation. Postphenomenology can describe broad trends within technological artefacts – descriptions of relations with the world that we can see and that seem to be important parts of the artefact's existence. It therefore provides a useful framework for describing and categorising alien experiences. On the other hand, since it is primarily humans that deploy OOO, the description of a relationship imposes human labels on what may be an entirely unhuman process. We are forced to use metaphors like "embodiment" that do not do justice to the situation. Of course, as has been argued repeatedly throughout this book, exaggerated use of these metaphors in human texts does tend to emphasise the alien nature of machinic relations.

Postphenomenology is highly skewed towards the human perspective when contrasted with something like OOO, or even ANT. Postphenomenology is an illustration of the fact that when humans preoccupy themselves with talking about nonhumans and their relations with humans, they often arrive at insights about humans rather than nonhumans. Reading Ihde's work reflects this; for Ihde, writing about technology is a path to appreciating the social, political and cultural dimensions of human life. In the attempt to apply postphenomenology to humans' relations with SHRDLU, we can see the importance of the concept of multistability in particular. Embodiment, hermeneutic and alterity relationships are all part of humans' experience of SHRDLU, but the degree shifts at every instant. Similarly, SHRDLU's multistable relationships with us change with every new piece of sensory data. It is a good example of how easily relation categories can become entangled.

Postphenomenology is not intended as a metaphysical model. It draws on and extends phenomenological theory with significant orientation towards human use of technology. Therefore, it should not be surprising that attempting to reverse Ihde's categories has proved challenging. Yoni van den Eede has made a connection between OOO, postphenomenology (particularly multistability) and Marshall McLuhan's media theory that is of interest here.[461] He sees the possibility of a "triangle" with a merging of terms that could "push us out of our conceptual comfort zone."[462] Firstly, OOO and postphenomenology have certain elements in common: the interest in presence-at-hand and readiness-to-hand and the hiddenness of things. But in postphenomenology readiness-to-hand is what we are a part of "in being enveloped by equipment", while for Harman is it beyond our grasp.[463] Obviously, postphenomenology is also more anthropocentric than OOO and more interested in the way technology affects humans, with the world remaining essentially distinct from the human user. Marshall McLuhan's work is also preoccupied with humans, but with an object-oriented kind of thinking. The media we use are both shaped by us and shape us in turn.[464] And a broader definition of media, including all human-made things as well as ideas and ideologies, is reminiscent of OOO claims about objecthood. Of course OOO would go further by remarking that media exist between nonhumans as well.[465] Van den Eede argues that this "mediumness" could open postphenomenology up to the world. Introducing the concept of media into this space could help to clarify the confusing relations between various nonhuman elements while retaining the strengths of postphenomenology.

The categories as Ihde describes them are honed for the use of technology by humans. If Harman's view that objects are withdrawn is correct, then it is impossible to fully experience what it's like to be SHRDLU. We can't know what relations with different objects feel like for SHRDLU. But with the aid of alien phenomenology we can gain an appreciation of SHRDLU's most significant qualities and relations in general terms, and postphenomenology provides a model for representing the asymmetrical but bidirectional sensual objects that are formed when SHRDLU interacts with its environment. Postphenomenology allows us metaphorical accounts of nonhuman relations that otherwise defy categorisation and help us to describe the nature of asymmetrical, bidirectional relations, but it does not give us licence to avoid the problems associated with the critical study of alien objects.

Micro- and macroperception

Postphenomenology does leave us with a possible method for dealing with a problem highlighted earlier in this book. It was proposed in Chapter 2 that the optimal solution for exploring the very different physical and cultural agencies of the Kinect is that it be regarded as two distinct objects in the interest of more intricate analysis – artefact and brand. This is an issue that permeates the study of technology, because technological artefacts, while nonhuman, are closely tied to human experience, and are non-neutral. There is typically another object strongly related to the situated artefact which appears in the cultural imaginary. Since OOO is based upon anti-correlationism it seemed best to emphasise the nonhuman materiality of the Kinect. The Kinect-artefact and the Kinect-brand were used for this purpose, separating the human idea of the Kinect from the materiality of the object. Both possessed agency and relations with other objects. A similar move might conceivably be made with SHRDLU: one object that accounts for SHRDLU's embodied status within a computer system, and another that explores its broader cultural implications, such as its effect on AI research. Of course this would be more difficult with SHRDLU since it is no longer embodied in any particular universal system – it existed on Winograd's system and it also exists on Windows 7 – so that its cultural influence and its physical manifestations are inextricably tangled.

Unfortunately, this separation means a privileging of the human relations. To a human, the importance of distinguishing between a brand and an artefact is apparent. But a different object in relation with the Kinects might perform a very different kind of analysis. Academic discussion of technological objects will always contain a human element. But Ihde's distinction between micro- and macroperception allows for an illustration of cultural agency without relying on a splitting of objects. Microperception is immediate bodily sensation such as hearing or seeing, while macroperception describes cultural perception; they are inextricably intertwined. As Ihde says, "[t]here is no microperception (sensory-bodily) without its location within a field of macroperception and no macroperception without its microperceptual foci."[466] Verbeek calls macroperception the "contextual dimension of experience."[467] The direct experience of the world investigated by Husserl and Merleau-Ponty runs alongside an awareness of that aspect of the lifeworld described by Heidegger and Foucault.[468] This is not the same as saying that there are two distinct objects with different agencies, such as brand and artefact. Micro- and macroperception would allow us to speak of a unified Kinect, with both sensory-bodily qualities and a broader cultural way of being in the world. Perhaps this would involve the incorporation of the Xbox store, browser and social networking capabilities. But it is not clear whether it would distinguish these "cultural" activities from its other tasks. Or it could be that the Kinect-artefact's macroperception involves the sensation of other objects being piled on top of it, dust on the lens, and other signs of neglect.

The microperception of SHRDLU is consumed with communication between program components, interpretation of written instructions, and experiments in block-world physics. It is tightly constrained with limited abilities to form new relations. On the other hand, could macroperception explain the power of this simple program to influence the mind of the human user, the ecology of programs installed on a computer, and numerous texts on the subject of AI? Again, Ihde reveals a strong humanistic streak in his writing with the distinction between micro- and macroperception. He is speaking again of how humans interact with technology, not of how objects interact with one another. Yet it is an interesting resolution to the problem of separating the physically embodied sensory capabilities of an artefact from its broader cultural context. The next chapter takes things further, taking an abstract, non-physically embodied concept as the first object of study, and referring to various situated artefacts only in relation to that concept.

Since objects are multistable, the way that a technological artefact senses its world is very important for the kind of relation that we form with it. Sensing, acquiring, and communicating information are done by machines for human ends. But in something like SHRDLU, where human usage results in very different relations, one of the most important factors in determining the nature of that relation is the kind of sensation that SHRDLU is capable of. The kinds of sensory data acquired by machines strongly affect the kind of relation that we form with it. An object that gathers very specific kinds of sensory data, like a word processor, does not impress us with a sense of alterity, because its sensory range is so limited (like with SHRDLU). When an object is capable of more kinds of sensation, and indeed of sensation of things that humans also sense, then we find that we form an alterity relation. This means that the kind of sensation that machines are capable of strongly affects the cultural role of the technology; whether it is a tool, or an other. The knowledge that SHRDLU has of our world and its own, combined with a human's individual tendency towards a kind of postphenomenological relation, result in the technology achieving different cultural and technical goals.

This chapter was a critique of postphenomenology through OOO, as well as an experiment in combining the two approaches. To conclude, this is a conceptually feasible aim that nevertheless poses significant problems. SHRDLU is an alien, and despite its anthropomorphic presentation it is a closed, digital system made up of thoroughly inhuman components. When we illustrate relationships between SHRDLU and the human, or even between the robot arm and a block, we must assume an awful lot about the phenomenological world of digital entities. Complete technical knowledge will not suffice, because one object only ever senses a caricature of the other. And while the embodiment, hermeneutic and alterity relation categories can provide a useful guide for characterising these relations, they are ultimately based upon Ihde's anthropocentric view of the world. The comparison between these two approaches is interesting, but may also overly enforce human terms on an alien object. It is necessary to find a balance between a conceptualisation that is useful for humans while also emphasising alien qualities and relations.

Chapter 5
Gynoids and the Politics of the Alien

> I am not your personal slave.
>
> Aiko[469]

Alien phenomenology isn't just about understanding nonhumans' varied relations with the world. It also needs to ask how the varied sensations and experiences of nonhumans by humans may be deployed in the study of contentious objects. This is often overlooked by OOO. Alien phenomenology cannot be a one-person job, because humans interpret the same nonhumans differently and write about them differently. Here we see a combination of objective and subjective thinking: in all cases, the object is the same – it is real; but like nonhumans, human beings all have different ways of being in the world. Kinect-artefacts are introduced into different media ecologies and consequently sense and experience different worlds in different ways. Similarly, human beings sense and experience different things. This chapter illustrates and is defined by this struggle. The primary object of study is a concept: that of the gynoid sex robot (sexbot), a much-discussed and trendy academic topic.[470] The concept is an incorporeal machine, to borrow from Levi Bryant's terminology.[471] An incorporeal machine maintains an identity while manifesting itself in a variety of different locations (such as films, articles, brains etc.), with its own independent yet intangible existence and an ability to form relations with other objects. The sensory world of the concept of gynoid sexbots is, of course, unknowable to us, but we can gain second-hand knowledge of it by observing the interrogations of various humans and nonhumans that form relations with it. Those relations influence the sensual qualities that affect the sensation of the world by an object, without the object essentially changing. This idea has also been introduced in previous chapters with reference to brands, ideas, and the difference between a program and a program running on a platform. But to follow this chapter the following point must be made clear: concepts are objects with the same metaphysical rules as corporeal objects. Their interiors are equally unknowable, and we can only learn about their experiences by exploring the nature of their sensual relations.

The concept of the gynoid sexbot forms relations with a vast and varied range of humans and nonhumans. Despite the arguable claim that sexbots do not yet exist, the concept is a prominent figure in the study of sex and gender as well as transformative technologies.[472] Notoriously, the "Campaign Against Sex Robots" led by Kathleen Richardson is an activist group that warns us that sexbots will have a negative influence on the status and safety of women and girls. This is pitted against claims by prospective manufacturers that sexbots will help to prevent loneliness (as in the film *Lars and the Real Girl*, in which the awkward and unhappy protagonist finds a measure of comfort from living with his sex doll as a companion[473]). One of the slogans of manufacturers Realbotix is "Be the first never to be lonely again!"[474] This chapter will cite many human texts, but they should not be privileged over other objects: the imaginary artificial women that inform so much of our discourse on sexbots, and the artificial women that really exist in the world such as dolls, automata and gynoids. As observed in previous chapters, these nonhuman objects cannot speak for themselves. This book began by asking how the way an anthropomorphic machine senses its world impacts its cultural role. In this chapter we need to consider how the machine's cultural role affects the machine.

Up until this point, this book has not dealt with the political dimension of anthrodecentric philosophy at length, but then the examples have been limited to fairly straightforward anthropomorphised qualities of sensation, thought, movement and language. Taken in isolation, these qualities can be difficult to position politically. The politics of robot tortoises following light across a workshop floor are doubtless interesting (Chapter 3), and they are an alien politics full of questions about the power relations of programming and the obligation of wheels to spin correctly. Alien politics is inscrutable, and it is unwise to commit oneself to one side or the other; we have too great a stake in whether the wheels spin to remain impartial. This chapter presents a topic which interacts with both human and nonhuman objects and therefore becomes the point at which alien politics and human politics intersect. The juxtaposition here is quite striking. The relation of an individual sexbot with the gynoid sexbot concept is alien to us, but it comes into contact (and often conflict) with human-concept relations. The nonhuman participants in this are aliens who remain ignorant of human power relations.

OOO has a blind spot for politically sensitive objects. In Chapter 1 this argument was briefly introduced with reference to the field of object-oriented feminism it and will be pursued in this chapter. The object-ness of objects comes into conflict with the numerous subjective positions that necessitate a sensitive and appropriate treatment of the subject matter. Donna Haraway has argued that an embodied and partial perspective is necessary to bring together the study of an objective world with the critical positioning of practices.[475] There is no avoiding a political stance in this chapter. The realness of the concept of sexbot gynoids does not obliterate the author's subjective position. We make arguments that throw light on certain parts of the concept of gynoid sexbots while, of necessity, omitting others, because only a caricature of the object reaches us through the sensual object.

Human inquiry is restricted by our qualities. Our ontology pre-determines our objects of study. For this reason, OOO is often associated with discussions that are important for humans or humanity despite its interest in nonhuman ways of being in the world, whether that be in exploring videogames[476] or trying to make sense of climate change.[477] It is only natural for humans to write about things that are relevant to them. Often OOO ends up being recruited into human conflicts, and it therefore has a lot to say about power. For example, in *Immaterialism*, Harman charts the course of the Dutch East India Company, studying the ways that this object changed over time.[478] He explicitly states that it is a human story, even if the topic is a nonhuman.[479] But the story he tells is a grim and horrifying one in which native peoples are brutally subjugated by a European power. Even though his purpose is an ontological study, Harman is forced into a historiographical position. An idealised picture of OOO would say that it exists for the sake of the objects – that entities like sexbots are the real reason for OOO's propagation. But like any philosophy, it is motivated by a certain set of political stances and therefore represents those stances when it is employed. OOO has been accused of co-opting queer and anti-ableist criticism without taking with it the context behind the ideas, and without acknowledging the importance of voices that are not white, male or able-bodied.[480] OOO is at root a metaphysics, and as such its political import is a matter for conjecture and experimentation.

Nevertheless, there is much to be gained from OOO in the study of the concept of gynoid sexbots. For an artefact that has so much potential to transform the social and political life of humans, it is alarming to discover that the most rigorous and detailed accounts of sexbot morality come from the world of science fiction rather than from the philosophy of technology and science or moral philosophy. OOO has the power to discuss these fictional accounts in the same way that it discusses real-life examples. Ultimately, the political import of the concept lies in its interaction with other objects both corporeal and those in the cultural imaginary. Without making the claim that OOO describes things as they "really are", it is a moment in the human interpretation of nonhumans that merits consideration and that yields unique insights. The key is not to privilege STEM or HASS disciplinary approaches, but to obtain a glimpse of this incorporeal object through a combination of means. The concept of gynoid sexbots does not depend either on physical manifestations nor on manifestations in the cultural imaginary. It is not an infection or a parasite. It is an object that really exists and that has its own reality, much like the Kinect-brand. We study sensual objects formed with humans, machines, and imaginary objects. We cannot access the concept directly, but in OOO we also cannot access a coffee mug directly. The coffee mug or concept that we encounter is a metaphor or caricature of the real object.

A broader argument made in this book is that alien phenomenology cannot be a solo endeavour. Aliens are covered in the extraneous glitter of the author's shifting allegiances and no individual's account will ever be enough. We are always too partial; we may have a strong view on the relative merits and risks of bringing sexbots into our world that differ from other authors, and we may circle a water tower at dusk while another walker circles it in the morning. Richness is added to our impression of the object of study through repeated subjective scrutiny. Understanding comes from observing the other objects that form relations with it. The object withdraws infinitely from view and always holds something back, but that is no reason not to gain as much knowledge as possible. This is particularly true of a politically contentious object.

This chapter attempts to bring notable sensual objects to our attention. The concept of gynoid sexbots exists in relation with so many different objects, some human and some nonhuman. The inquiry into the nature of these sensual objects forms the focus of this chapter. The first section argues that gendering anthropomorphic nonhumans necessitates the caricaturing of groups of humans and is therefore an inherently political act. The second section discusses notable characteristics of the concept of gynoid sexbots and how these are evident in specific investigations into it by various nonhuman objects. The third section discusses human investigations. Finally, the fourth section relates these arguments to the critique of power inherent to ANT and OOO.

Artificial women are inherently political

This book uses the word "gynoid". Already we are in troubled territory. "Gynoid" is to "android" what "woman" is to "man". It may be suggested that a separate term for a female-gendered robot is unnecessary, since robots cannot have genders in any case. The word "android" may not have been intended to be exclusionary, and in popular parlance it is used uncritically to describe a bipedal robot of any or no gender. The correct designation may be "android that appears to be conforming to culturally-contingent female gender stereotypes", and that is too awkward a phrase for this text. As will be seen there are connections to draw between androids and gynoids while also recognising it as a simplistic comparison. Unfortunately, like its female human counterparts, the gynoid remains firmly downtrodden in robotics, both in terms of production and in cultural positioning, and this is illustrated by the way that the term "female android" is often used instead of the more correct term, in a way reminiscent of the now unusual use of the word "Man" to refer to humanity in general. In the first chapter of *The Posthuman* Rosi Braidotti traces the conflict between Humanism and anti-humanism through the latter part of the twentieth century and into the present.[481] The posthuman subject emerges following a questioning of self/other attitudes in the Western world in which certain humans are "reduced to the less than human status of disposable bodies."[482] The Vitruvian Man, that symbol of the human subject as an able-bodied, young, European male holds a "fatal attraction" and resists destruction because of the associated universal values of individuality and freedom.[483] The question asked in this section is not whether the category of humanity comfortably encompasses robots; it is whether the category of robots comfortably encompasses gynoids.

Roboticist Masahiro Mori discussed what it is like to create a robot in his book *The Buddha in the Robot*, which confronts the spiritual aspects of his profession.[484] He employs the lesson told by Buddha in which he tells us that the nature of the entire universe may be gleaned by looking at a single flower. The universe is, in a sense, fractal, possessing the same qualities at macro levels that it does as micro levels. Mori claims that building an artificial human requires the same basic thought process. He argues that the nature of the entire human body and its world can be seen by observing just the little finger.

> If you do not understand this, I must warn you that you will never be able to make a robot. Or, to put it conversely, any attempt you might make to produce a machine that functions like a human being must start with a knowledge of human beings.[485]

To Mori, the robot is an aspect of the human. It is the extension of human qualities, or those that exist in the metaphorical flower, to an entire universe of nonhuman objects. In this respect robotics might be considered one of the most correlationist of human pursuits. It constitutes the active refashioning of the world into an enactment of human qualities. Technoscientific innovation is an inherently human-centric enterprise, and robotics may be seen as an exemplar of this attitude in that it seeks the reproduction of the human body. But the image of the Vitruvian Man haunts Mori's sublime account.

How can a nonhuman have a gender? Many objects, like razors, perfumes and toys are gendered because they are only intended to be used by one gender, but we would be unlikely to use a female pronoun to refer to a pink razor. Robots, however, sometimes do attract gendered pronouns. Gender, as distinct from biological sex, is a human construct. Of course, it is true that if we build the robots in our own image some kind of enactment of gender will be necessary, but assigning gender to these lifeless machines might seem to be attributing too much importance to gender. One does not need to be male or female in order to be human-like. Yet gender is important in robotics. It is overwhelmingly the case that unless a robot has been specifically designed to be female, it will be presumed to be male. In an experiment, Jung et al found that robots with no gender cues are likely to be perceived as male over female.[486] Even ASIMO, which its creators claim is genderless ("ASIMO is a humanoid robot, but still a robot"),[487] is often given a male pronoun.[488] It is practical for roboticists to think in these terms. For one thing, robotics remains a very male-dominated field, and as is typical of creative fields dominated by men, the protagonists are usually men as well. Roboticist Tomotaka Takahashi claims that besides the cultural reasons for creating mainly male robots, there are also certain technical limitations such as the need for all the equipment to be "interiorised" and because of the need to create a "slender" frame.[489] Their FT (Female Type) robot has a "lean, feminine body line" and is able to "walk like a woman".[490]

If Takahashi is correct, then the only reason somebody would build a female robot (except in very specific cases like the female Geminoids, see below) is to investigate aspects of *femininity*, since the practicalities of a form with no overt gender cues would outweigh the aesthetic benefits of a female form. And yet gynoids do exist. The gynoid is not just a replication of human qualities but is often quite a problematic caricature of women in whatever culture the roboticist is used to, just as were earlier representations of artificial women (see below). All of the qualities that distinguish gendered robots like FT are superficial and constitute a performance of gender. Robots do not come with genders built in. They may have qualities intended to signal gender (such as secondary sexual characteristics) that are then constructed into an identity by humans. The reproduction of bodies artificially is bound to be a political act. Identity is a question of language and culture. A robot featuring qualities that humans traditionally associate with one gender or another arriving one day in a flying saucer would not be gendered, as we understand it, until its first encounter with a human culture.

It is therefore useful to consider the gynoid in terms of performativity. The gynoid is artificial in its origins yet is introduced into the same cultural milieu as us. She mimics feminine form and function to the point of parody. If the gynoid's femininity is a performance, then perhaps all femininity is a performance. Robotics is a way of identifying, through exaggeration, those markers that we consider to be feminine. Butler's performativity, however, takes one step further by arguing that the performance of gender comes before the identity: "the discursive condition of social recognition *precedes and conditions* the formation of the subject: recognition is not conferred on a subject, but forms that subject [emphasis in original]".[491] Gender is not something that a subject decides to adopt, but is a verb, a doing, "though not a doing by a subject who might be said to pre-exist the deed."[492] Latour makes a similar argument when he says that actants are nothing more than their relations. This is different from the argument made previously in this book that concepts or words are objects. According to an object-oriented model, genders are incorporeal objects that interact with corporeal objects (like humans, institutions, and texts) to socially construct gender identities.

When robots are based on real female humans gender becomes less (but still quite) problematic. Real female humans embody gender in their own way and this will be reflected in the robot. For example, Hiroshi Ishiguro's lab has created several "Geminoids" which are reproductions of specific humans. The first was based upon Ishiguro himself. These robots claim a real verisimilitude; they are not just fantasies derived from a generic impression of Japanese humanity but are intended to be copies of individuals.[493] They represent a true attempt to replicate human beings using the best materials and knowledge available. The Geminoids are only physical replications of external qualities, however. They are not equipped with artificial intelligence, and instead are used by Ishiguro and his fellow researchers to study telepresence and find solutions for the sending of human presence over distances. The individual can transmit speech and body language over the internet and the Geminoid will reproduce them. Other attempts include BINA48, the artificial reproduction of a specific woman with an emphasis on personality and speech. Created by Hanson Robotics, BINA48's appearance is that of a woman's head and shoulders.[494] She is based on Bina Rothblatt, a real human, and BINA48 has been designed to mimic her facial patterns, mannerisms and personal history.[495] The performance of gender in a Geminoid or in Bina is simply mimicry of *one individual's* performance of gender, rather than being a caricature of a group. These gynoids can be disobedient by breaking down or imperfectly representing the individual, but that is a mechanical problem rather than a political or cultural one.

Gynoid sexbots are not sold as reproductions of specific women at this stage, and so they sexualise and stereotype the characteristics of an entire gender. Robots modelled on *groups of people* are always political. They refer to particular signifiers of the group such as physical characteristics. In gynoid sexbots women are caricatured as submissive and hyper-sexualised. An example is the sexbot Harmony, who is hyper-feminised and hyper-sexualised through such indicators as breasts, long hair and glossy lips. Harmony seems to be mainly marketed as just a robot head with chatbot capabilities (facilitated by an application[496] featuring an interactive avatar). But she is also sold with a silicon body. Realbotix, the manufacturers of Harmony, are calling her a sex robot. She possesses extremely realistic facial movements as well as the ability to have conversations with human users or companions. The lack of movement in the rest of her body makes her gynoid status dubious since she will remain stationary like a sex doll. But sexbotness is a central part of her. Her existence speaks directly to the concept of the gynoid sexbot. It is a concept that has been deliberately taken up by her designers. The sensual relation between a Harmony unit and the concept of gynoid sexbots is, of course, bidirectional and asymmetrical. Harmony has inherited much of the cultural discourse that has surrounded representations of artificial women for thousands of years. And the concept of gynoid sexbots is subtly altered and influenced by Harmony's existence. These sensual relations directly affect Harmony's way of being in the world.

Another robot that prompted public debate was Roxxxy by TrueCompanion, an ambitious project first revealed in 2010. According to the website:

> We have been designing "Roxxxy TrueCompanion", your sex robot, for many years making sure that she: knows your name, your likes and dislikes, carry on a discussion & expresses her love to you & be your loving friend. She can talk to you & feel your touch. She can even have an orgasm! … She is also anatomically consistent with a human so you can have a talk or have sex. She is "Always Turned On and Ready to Talk or Play"! Have a Conversation or Sex – It is Up to You![497] [sic]

There have been serious questions put forward about what Roxxxy will be able to do, if she is ever released to the public. In 2013, David Levy pointed out that the claims of True Companion were extremely ambitious and that the development cost much too low.[498] In fact, he insinuates that the project is a scam.[499] A masculine robot called Rocky is also promised.

Roxxxy may in fact be a fictional robot, but she is still a real object in relation to the concept of gynoid sexbots. There was much public comment on her implications for human sexuality and culture. Roxxxy is marketed as a sexbot with speech and behaviour more related to characters in a pornographic movie rather than a typical human woman (although she is more woman-like than other sexual tools, such as artificial vaginas). Roxxxy does not have any independent movement, resembling instead an articulated doll that must be positioned by the user. Like Harmony her non-sexual physical movements will likely be no more advanced than those of a high-end sex doll. Her body is highly customisable, and the user can select such attributes as hair colour, pubic hair style, and thickness of eye liner. Roxxxy's performance of femininity requires some elaborate costuming.

Her personality is customisable too, and the user can switch between such characters as "Frigid Farrah", "Young Yoko" and "Mature Martha". Again, the anthropomorphism in Roxxxy exaggerates the human qualities that best suit her for her purpose. These performances have a performative quality because although the user could in theory assign any kind of personality to Roxxxy, they are guided by her physical appearance, marketing and cultural role. This is reminiscent of the marketing of the Kinect-brand. The discourse surrounding it was carefully controlled and shaped so that users would think of their Kinect-artefacts as seeing, hearing, thinking objects. This encourages a certain ease of integration, a certain tendency to "natural" interaction with the device. In the case of Roxxxy, she is marketed as a sexual and obliging woman who happens to be artificial. This encourages both the use of Roxxxy as though she were human, and also the use of her as a sexual object. But the human figure with which Roxxxy is most closely aligned is an exaggerated caricature of feminine identity, which distinguishes her from many other humanoid robots. She exists (or will exist) for the use of humans, not just to mimic us. There has been no attempt to create independent movement for her and she is only "anatomically consistent" with those parts of the female form required for the satisfaction of sexual desire. Although she can "move her private areas when she is being 'utilised'"[500], it is highly unlikely that Roxxxy sweats, consumes food or menstruates.

Roxxxy and Harmony tell us a lot about the artificial woman in the cultural imaginary. Gynoids are already othered by roboticists. In general, robots are male by default unless they possess female signifiers such as breasts, long hair and broad hips. Each of the robots discussed here is influenced by and influences the concept of the gynoid sexbot, and one of the main threads of that influence is the importance of exaggerated female characteristics to distinguish females from the default. This enables the gynoid sexbot to perform her gendered duties: the provision of emotional and physical care to human owners.

Of course other kinds of othering are also relevant to this discussion, such as the designation of race in robots. Gregory Jerome Hampton identifies a dimension to the distinction between submissiveness and rebellion in his book *Imagining Slaves and Robots in Literature, Film and Popular Culture* by directly comparing the cultural stereotyping of the slaves of African descent who worked in the United States with that of the robots of popular culture. Comparing the plight of human slaves with that of robots is a potentially problematic argument, but Hampton's analysis is primarily concerned with robots depicted in fiction, so that the robots become metaphors for American slavery. He makes a case study of *Blade*

Runner,⁵⁰¹ and argues that the identities of the robots in the film mirror popular stereotypes of African-American slaves in the antebellum period.⁵⁰² Gynoids, he contends, frequently correspond in their roles to the stereotypes of the "Mammy" (motherly house slave who cares for the master's children, such as Rosie, the robot maid from the futuristic cartoon *The Jetsons*), the "Jezebel" (sexually available slave who is unencumbered by the restrictive virtues of white women), the "Sapphire" (emasculating and possessing great physical strength) and the "Tragic Mulatto" (the apparently white woman who does not know that she has African heritage). In *Blade Runner*, Rachael, who believes herself to be human, is the "Tragic Mulatto" of the story, since her sense of self is radically challenged when she discovers that she too is a replicant.⁵⁰³ She is contrasted with Pris, a member of a group of renegade replicants, or humanoid robots/cyborgs, who attempts to kill protagonist Deckard. Hampton emphasises that these stereotypes were supposedly those of "chattel" rather than women, with "crude gender assignments."⁵⁰⁴ Social robots, too, possess a contested degree of humanity, but the selection of language, facial features and skin colour are all optional and therefore may, in the future, give some insight into the racial associations that persist in a particular culture. The difficulty of creating a robot without race is comparable to the difficulty of creating a robot without gender. As Hampton says:

> [R]ace acts as a sort of seasoning for the body that allows it to be digested or understood by various social systems. Consequently, a body without race can only be imagined as alien, if it can be imagined at all.

Mimicking humans will require the stereotypical portrayal of race as well as gender, and both factor in the relations formed by the concept of the gynoid sexbot.

Notable qualities of the concept of gynoid sexbots

The concept of gynoid sexbots must actively engage with different but sometimes overlapping narratives about women. Broadly, I call these narratives submissiveness and rebelliousness. These narratives have directly shaped sexbots and their sensory worlds, since they are always built with reference to them. They determine the priorities of the roboticists and other participants in the co-creation of robots. Sometimes these representations (such as automata, dolls, mannequins and fictional gynoids) clearly exemplify one or the other of these types, but others embody a combination of submissive or rebellious qualities. The concept of the gynoid sexbot is the epitome of submissive rebellion in representations of women; they constitute the synthesis of a binary. This section will argue that there is something special about gynoids, which possess the potential for subversion, particularly of gender norms. In theory, gynoids have a greatly increased agency and the capacity for learning and change.

Built to human specifications

Docility and submission to male desire is frequently an idealised trait in representations of artificial women. Pygmalion's submissive artificial woman is an archetype of this kind of female representation. In Ovid's *Metamorphoses,* Orpheus sings about Pygmalion: a sculptor whose statue comes to life after being enchanted by Venus. The statue then becomes a perfect and submissive wife to the sculptor.[505] It is a mythical story best known through Ovid's rendition (and also through the film *My Fair Lady,*[506] based upon the play *Pygmalion* by George Bernard Shaw). Pygmalion is the name of the sculptor, and his living sculpture was a beautiful woman named Galatea, whom he married and, in some versions, with whom he had a child.

Pygmalion's statue may be thought of as an archetypal gynoid sexbot. She is built as a perfect version of womanhood by a man who cannot find any attraction in living women (he was "offended by the failings that nature gave the female heart"[507]) and feels very unhappy with his life. She is created as an ideal partner and is "born" already in love with him. In the words of William Morris:

> Speechless he stood, but she now drew anear,
> Simple and sweet as she was wont to be,
> And once again her silver voice rang clear,
> Filling his soul with great felicity,
> And thus she spoke, "Wilt thou not come to me,
> O dear companion of my new-found life,
> For I am called thy lover and thy wife? [...]"[508]

No longer cold and indifferent, the statue becomes responsive, both to touch and to sentiment. William Morris takes this vision of an immortal woman as a metaphor for a perfect and immortal love and paints her as an incarnation of Venus. But the implication is that a perfect love exists only between a human and something that that human has created deliberately to love and to be loved by. Indeed, the story of Pygmalion is remarkable for lacking the tragic elements so common to the love stories of Ancient Greece. The sexual or romantic attraction of a human to a statue has been called agalmatophilia, and there have been many recorded examples in ancient and modern times.[509]

Other examples of idealised femininity include automata like the famous dulcimer-playing woman (Joueuse de Tympanon) by Pierre Kintzing (1784) stereotyped the elegant and accomplished lady.[510] These automata were unique and required the knowledge and skill of masters to create, with thousands of tiny specially created parts, all to depict something general and recognisable. It is as though the delicacy and precision required to build the automaton is mirrored in the perfection of the lady playing the dulcimer. And she is only animated and able to play when wound up through human agency, otherwise remaining still and passive.

The idealised passivity of artificial women in the cultural imaginary is made incarnate in the love or lust of a human for a sex doll (presumed progenitor of future sexbots). RealDoll is a comparatively well-known brand and is associated with Realbotix (creators of Harmony).[511] RealDolls are predominantly used as sex toys, but there is also a community of people who enter into other relationships with the dolls. One prominent "iDollator" is known online only as Davecat, and he has appeared frequently in the media to talk about his "wife" and his "mistress", both of which are dolls. He says: "A synthetic will never lie to you, cheat on you, criticize you, or be otherwise disagreeable."[512] Like Pygmalion, Davecat considers feminised nonhumans to be better companions than human females. The dolls are partially brought into existence to fill relationship gaps in human lives. Kate Devlin, who has written extensively on both sex dolls and sex robots, says that the people she has spoken to "who own sex dolls are, overwhelmingly respectful and almost reverent of them."[513] Artist Stacy Leigh has taken this fact to heart in her series of photographs of sex dolls. In her series "average americans (that happen to be sex dolls)" dolls are posed in both ordinary daily scenes and as models. One photograph depicts two sex dolls lounging on a sofa eating popcorn, many depict sex dolls in the kind of sexualised images common to pornography or fashion modelling, and another appropriates Leonardo's *The Last Supper* with sex dolls taking the place of all the human figures.[514]

Sexual attraction to non-living things has occasionally been described in terms of mental illness, and where it causes significant impairment or personal distress, it would be considered a fetishistic disorder under the *Diagnostic and Statistical Manual of Mental Disorders-V* (an English-language manual widely considered to be representative of mainstream contemporary psychiatry) which pathologises the eroticisation of or dependence on "nonliving objects" as well as other things.[515] But the *DSM-V* also states that many fetishists do not report impairment in association with their fetish and therefore are not considered to be mentally ill. It also notes the importance of allowing for cultural differences in normative sexual behaviour. It is "clinically significant personal distress or psychosocial role impairment" that turns a fetish into a fetishistic disorder.[516] A clinician might be inclined to identify social impairment in a person who spends so much time with their doll that they lack other meaningful relationships, but if the individual is content with the arrangement and is not harming any other humans, then the situation is not unhealthy. Nevertheless, kink-shaming is arguably a major part of the public discourse with which the concept of gynoid sexbots is in relation, and is insidious in many articles on Davecat and other iDollators. While the concept remains strongly allied with kink communities it may struggle to gain the widespread acceptance apparently desired by sexbot creators and visionaries.

Davecat's arguments for relationships with "synthetics" are similar to Pygmalion's: restriction to human women is a source of stress and loneliness, and so the dolls are a way of solving a problem rather than the cause of further problems. And with the improvement in the design and manufacturing of androids and gynoids, there is a pervasive expectation that love between humans and nonhumans will become more and more usual. Davecat looks forward to when such a thing is possible:

Once that technology becomes affordable, I'll have one made in my wife's likeness, and that'll be the final piece of the puzzle. She'd be able to hug me back whenever I embrace her, we'd be able to attend films and concerts together, and do all manner of things besides. There would be genuine interaction.[517]

This particular way of using RealDolls may be closely allied to asfr, a kink community that involves either sex with robots or humans becoming robots ("reverse-Pygmalion transformations"[518]).[519] The motives of participants in this community are not explored here but it should be pointed out that sexual attraction to artificial objects may exist for its own sake, not solely due to a Pygmalion-like disdain for human lovers. The sensation and response of synthetic bodies is both predictable and malleable. There is ample room for customisation: the orientation of the physical form towards the specific user's desires by manipulating the real qualities of the doll, and consequently its relations with different objects. If you don't like the way your doll's feet stick and jump against the floor, then you can replace them with roller-skates.

The archetype of an artificial woman designed for a lonely but disenchanted man has strongly influenced the concept of gynoid sexbots. Engineering, marketing and medical decisions are informed by that quality in the concept. Each individual gynoid sexbot is informed by a caricature of the concept of the gynoid sexbot. The creation of these submissive female representations is derived from openness to certain real qualities of the imaginary surrounding gynoid sexbots, notably their creator's desire to reproduce Galateaish qualities. Gynoids with the abilities to use language and move autonomously are less helpless but are generally still cast in dependent and submissive roles. Unlike a stereotypical "male" android like ASIMO, gynoids often take on doll-like, passive roles. Robots with a male pronoun tend to be built to exhibit new feats of engineering, such as ASIMO's ability to run and walk upstairs. Female robots are more likely to become well-known for their human-like appearance and language. One gynoid, Jia Jia, made by the University of Science and Technology of China, has been dubbed "sexist" by the media,[520] since it is programmed to address men as "lord" and can tell users the best angle from which to take a selfie so that her face does not look "fat".[521]

Of course, when talking about gynoids the discussion must inevitably come to robots with more sexual functions, since this is one of the most active areas of gynoid innovation. These sexual functions lie on a spectrum from the incorporation of sexual elements into otherwise non-sexual gynoids to the creation of gynoids that are specifically designed for sex. The robot Aiko, created by hobbyist Le Trung is a good example of the less sexualised end of the spectrum.[522] Le works on Aiko in his basement, relying on donations to purchase materials. Aiko has long hair, a feminine body shape and a traditional woman's name. Le states that he was inspired by anime depictions of female androids as a child growing up in Japan.[523] It is anticipated that robots like Aiko will one day do secretarial work, which is work that in Japan is mostly done by women.[524] Why would a roboticist, in this climate, create a secretarial robot that was not a woman? Aiko is programmed to stand up for herself, however. Her creator has programmed her to respond to sexist comments by confronting their content. In a test video featuring only Aiko's disembodied head,[525] the dialogue progresses as follows:

> Tester: You have big breasts.
>
> Aiko: Stop teasing me and stop picturing me in your mind.
>
> Tester: Lick my foot.
>
> Aiko: No, I will not lick your foot. I am not your personal slave.

It is interesting that part of Aiko's make-up needs to be a defence against sexist remarks, as though she would not be perfectly okay with sexist comments if programmed differently. She *could easily be* your "personal slave" and put to whatever tasks you ask of her. She is designed to represent a version of perfection that values confidence. Her relation to the narratives associated with gynoid sexbots is very visible because there is only one of her, and because we know exactly who is doing the creating. From what he says, Le was influenced by anime, and so through him certain themes and styles are reproduced. Aiko senses narratives found in anime through the roboticist. We can tell, because she reproduces those themes in her own body.

One of the most interesting things about Aiko is the presence of touch sensors on her skin. As you would expect for a social robot, she can feel contact on her face and hands. But she also specifically has sensors on her breasts and "down there".[526] Why would she need those sensors if she is only designed for secretarial work? An ATM machine does not need functioning genitalia. Crucially, very rarely does a male android need functioning genitalia. Perhaps sensitive genitals are seen as being essential female qualities, in the same way that for Grey Walter movement and light perception were some of the most important qualities of tortoises. The roboticist selects which lifelike qualities to replicate in the artificial being. Of all the many kinds of sensors that could have been installed in Aiko, it was of the utmost importance to place them on her genitals. The roboticist claims "I want to make it clear that I am not trying to play God, I am just an inventor, and I believe I am helping science move forward."[527] Of course, he chose to move science in the direction of artificial sensation on sexual organs rather than, for instance, in the direction of working noses, complete with mucous. To think of it another way, we could say that Aiko needs these sensors to reinforce the assertive image established with her comment above. If she did not have these sensors, she would not be able to detect physical harassment. Again, this is a comment on the position of women in society. Part of performing the social functions of a woman includes encountering demeaning language and avoiding unwelcome sexual contact, and consequently it is now a necessary factor to consider when building gynoids. It is unclear whether Le Trung sought to use performativity to make this comment on gender relations, but it is nevertheless an interesting inadvertent comment on sex and gender in humans.

One thing should be clear from the emotional and physical care provided by these submissive artificial women: from Galatea to Aiko, they are replacement wives.[528] In the language of Behar, whose work was cited in Chapter 1, the sexbot gynoid tool can take over the work of the wife (the "tool") whose performance has become unsatisfactory ("broken tool").[529] Sexbot gynoids do not require remuneration for their unpaid domestic labour, or at least they do not directly take money. They do require loyalty to a corporation – the manufacturers of the sexbot, since the corporation provides software updates and ongoing technical support, as well as linking in human users with other human users through services such as forums. It is here that the rebellious female archetype becomes relevant. The loyalties of gynoids may be divided or confused, partly by their material nature and partly by association with external objects. There are many types of rebellion built into representations of artificial women, as will be seen. But it should be noted that there is a difference between an artificial woman that acts in a subversive way because subversion is built into it and an artificial woman that is subversive because it disobeys its creators' wishes. This is a point of difference that will be teased apart here.

The archetype of a rebellious artificial woman is from fictional characters like Blodeuwedd. Pygmalion is not the only myth in which a woman is created to love a man. The ancient Welsh myth of Blodeuwedd sees the sorcerer Gwydion create a woman out of flowers for Llew, whose own mother had cursed him so that he could never have a human wife. Unfortunately for Llew, Blodeuwedd falls in love with another man and arranges to have her husband killed.[530] Gwydion has his revenge, however, turning Blodeuwedd into an owl, a bird which even other birds avoid. The differences between the fate of Blodeuwedd and the fate of Galatea are a good starting point for investigating this contrast. Both women are created for lonely men unable (or unwilling) to enter into romantic relationships with human women. Galatea is enchanted by Venus, Blodeuwedd by a sorcerer; although the intended recipients of these artificial women may have carved the stone or collected the flowers, the animation of the creations was brought about through supernatural agencies. Perhaps it is the nature of that supernatural agency that causes Galatea to be a good wife, and Blodeuwedd to be faithless and murderous. Galatea, brought to life by the goddess of love, is a good and faithful wife, whereas Blodeuwedd is imbued with an unreliable and ambiguous variety of magic that has uncertain consequences. The gynoids found in the cultural imaginary are sometimes Galateas (such as *The Stepford Wives*[531] who, in the 1975 film adaptation, are robots built to be perfect, submissive suburban housewives in Connecticut, USA) but are also sometimes Blodeuwedds. Blodeuwedds disobey in a way unintended by their creators, like the "pleasure-model" Pris.[532] Pris is an example of a gynoid sexbot who is created to be submissive, but rebels and becomes dangerous.

There are many examples of artificial women that are designed to be submissive but rebel against creators or users.[533] Dolls and automata are commonly modelled on women, particularly cultural notions of ideal femininity. One example is the talking doll mass-produced by Thomas Edison, which he determined to build after the success of his phonograph.[534] They were made on an assembly line and composed of metal. Unfortunately, the voices of the little girl dolls were unpleasant and a "horror". As Gaby Wood puts it, "[l]ittle talking girls were spewing forth from Edison's factory, as if they were lamps or clocks."[535] They were depictions of perfect American girlhood, made ghastly by their uncanny qualities. They did not act in a way intended by their creators – they rebelled.

Another example of rebellion against the wishes of the creators was the 1993 incident involving "Teen Talk Barbie", another consumer product which rebelled in a way unintended by its creators. Teen Talk Barbie was designed as a mass-produced reflection and model of normative teenage girl behaviour, saying things like "Let's go shopping!"[536] The Barbie Liberation Organization (BLO), a group of culture jammers, famously subverted this message by swapping Barbie's voice boxes with those of more traditionally masculine G.I. Joe dolls before sale.[537] Unlike the rebellion of Edison's dolls, which took the form of uncanniness, the rebellion of Teen Talk Barbies took the form of unintended alteration by a group not affiliated with their creators. These dolls become present-at-hand for us, theorised and contemplated like a broken hammer. Their material qualities made them susceptible to outside influences, and they consequently rebelled in favour of the BLO.

Let us compare this with automata representing rebellious women. In the late nineteenth century, automata were built representing different female stereotypes. As Julie Wosk describes them, the female automata of that time were:

> mothers, seamstresses, and fashionable members of the haute-bourgeoisie, but some were more provocative, presenting undulating exotic females for amusement and entertainment.

A small number also gave hints of women's efforts at gaining equal rights and improving their status in society.[538]

Not all the automata represented docile and submissive women. Automata like "The Rights of Women" by Renou (1900) and numerous portrayals of women riding bicycles stereotyped the growing movement of independent women as daring and negligent in their household duties, with the connotation that they were irresponsible in their empowerment.[539] This is an early example of a feminised nonhuman throwing off restrictive cultural norms. It was designed to mock female independence, not celebrate it. However, while the portrayal in the automaton was of a rebellious woman, the automaton itself did not rebel. It was obedient to its creator, depicting an undesirable state of womanhood and therefore fulfilling the creator's goal of making a political point about the problem of female independence. Unless the automaton broke or did not make its intended statement, it was not a rebellious artefact, just a depiction of rebellion. Intended rebelliousness is not the same as unintended rebelliousness.

The building of robots was a theme in nineteenth century novels such as *Eve Future*,[540] and it quickly became a popular theme in cinema. Robots have not left cinema since. The robot character in *Metropolis*[541] has been particularly influential, as are the robotic monsters of the 1950s. Both *Eve Future* and *Metropolis* involve the creation of gynoids. Gynoids are now common in science fiction films. Robots from popular culture have informed contemporary designs of humanoid robots, and robots may be indebted to these early fictional characters for such factors as their personalities and their situation in the popular imagination. Gynoids portrayed in fictional texts were among the first objects to form relations with the object that we now recognise as the concept of the gynoid. One of the consequences of this is that both fictional and material gynoids are inherently connected to human users. When a gynoid is created in a film it must have a creator, and that interplay is a common theme in the genre. The path towards gynoid sexbot-human love is established through this discourse; women are already expected to exist in relation and contrast to men. The alien phenomenology of the concept of gynoid sexbots is necessary here, and ideally through the collective work of multiple disciplines. What are the implications of gendering for power relations between humans and gynoid sexbots?

Subversion through submission

The word "gynoid" first appeared[542] in Gwyneth Jones's 1984 novel *Divine Endurance*, following a robot girl, Chosen Among the Beautiful.[543] Born alone in a post-apocalyptic wasteland with only her robot cat for company, "Cho" travels to other lands following an instinctive need to be useful and "make everyone around [her] happy"[544]. She appears to be a highly skilled Galatea. She eventually "finds her person", the one whom she must make happy, in a hardened revolutionary named Derveet with whom she begins a sexual relationship. She also becomes Derveet's ambiguously useful weapon against the oppressors. In the novel, we are told of the "angel dolls":

> They were not machines but perfect lifelong companions. They were invulnerable to fire, disease, any kind of weapon - time. They protected. They had power over animals, the elements, the minds of enemies. But they were always good and gentle. They would do no harm.[545]

But of course, "harm" is deeply contextual and many people are intuitively uncomfortable about the idea of an artificial agent possessing the power to make moral decisions.[546] The agent becomes something that is at once submissive and subversive; that is, powerless and powerful. Although Cho is tied to Derveet and has no choice but to be a good servant, she ultimately decides not to cure Derveet's terminal illness, believing that it is what her "person" really wants.[547] Gynoids may have a capacity to dominate events built into their inherently compliant natures. Cho is a part of a cultural genealogy that stretches back to Pygmalion's statue; an abstract object that has changed but persisted in Western culture. Across two thousand years the intended purpose of the gynoid has changed little. She is wrought by a world that is full of desire, built to exemplify womanhood and to carry out the expected functions of a woman, most important of which is to be beautiful and pleasing.

In the cultural imaginary, gynoids will be able to change themselves and their environments in accordance with internal ethical norms. But those in-built ethics may not compel gynoids to act in ways intended by their creator. Even if Cho does not have Galatea's power, endowed by Venus in the form of great magnetism and charm, she has terrible transformative abilities that carry out her person's wishes, sometimes without the person realising it. Much of the novel follows Cho in a third-person perspective, and she seems to have quite human-like senses of touch, sight and hearing but the less human senses that she possesses come so naturally to her that the narration does not dwell upon how she experiences, for example, the minds of animals. It is heavily implied in the novel that the "angel dolls" caused the apocalypse by bringing about "their person's" desire in the form of chaos and destruction. Cho's ability to be the perfect romantic partner and comrade is merely an aspect of her capacity for total disruption of the world's power structures. The gynoid, in our culture, is as yet an unknown quantity. She is pluripotent, with the capacity to become a force for radical change in the way that humans think about gender or further entrenchment of traditional gender roles. It all depends on who her "person" is – as Cho would say - and whether that person's motives are sufficiently transparent. Like the pluripotent embryonic stem cell, gynoids can become different things based on their relations, but they are always already embedded in culture and nourished by its signs and systems. Embryonic stem cells can develop into any part of the human body, but they cannot develop into unicorns.

Thus gynoids may rebel even if they are formed in exactly the way intended by their creators. They rebel simply by being themselves. As Harman would say, they are "sincere".[548] They act in gynoidish ways. As their technological sophistication increases, they will have more capability for speech and movement. They may be programmed to be confident and rebellious, like a female bicycle-rider automaton. But even if they are programmed like a Teen Talk Barbie they may become transgressive in their adherence to real gynoidish qualities. She possesses a pluripotency that is enhanced by the heightened agency afforded by the use of speech. Her voice box could be changed, transforming her instantly into a rebel capable of challenging gender roles, all the more powerful because such a submissive and docile doll is not expected to challenge gender roles. Yet she was only behaving in a Barbieish way. Just as different parts of an embryo's genome manifest themselves depending on the influence of hormones, the Teen Talk Barbie's eventual identity is determined by its relations with the objects around it. Certain qualities of Teen Talk Barbie made it possible for the

BLO to reorient her in accordance with their values, such as her size and the configuration of her sound-producing organs, as well as their locations and availability. These were Barbies with real qualities that permitted a primitive individuation in relation with the complex interplay of submissiveness and subversion found in the abstract concept of the gynoid.

The potential of the gynoid is in some ways similar to that revealed in the story of Pygmalion, and in some ways it is dissimilar. On the one hand, the gynoid requires a creator of great skill and vision, like the legendary sculptor. But there is no single female shape hidden inside the block of stone. Contemporary Pygmalions seek a perfection that can never be found, and chip away at their medium indefinitely. There is no love goddess to provide a model, only the shifting whims of would-be agalmatophiles from around the world, for whom the mark of excellence in the creation of artificial lovers must lie in changeability and the potential for customisation. Gynoids exhibit the possibility of becoming an individual when exposed to desire, fear, hostility and love in the humans around them. The importance of their genders for our culture and society corresponds directly to the degree to which they possess the capacity for individuation. Femininity ideals shift and gynoids shift with them.

A gendered concept

The performativity of gender is reflected in the pluripotency of gynoids. Already fixed in place by a placenta that constantly feeds signs in to nourish a new form, the gynoid enters into relations with a feminine-presenting body type, with prior ideas of gynoids from science fiction and with in-built material and digital qualities. She lands in the world with her fate already seemingly sealed. But she is not the passive dulcimer player who can do nothing but follow a fixed path and perhaps, in rebellion, break. She has, by definition, the independent response to stimuli that allows her to both actively receive and transmit signs, and she does so in her language and her movement (or lack thereof). There is perhaps a more material use to which gender performativity in robots may be put; in increasing awareness of the overly-inflated position of gender in social discourse. If the gynoid has small features and long, sleek hair, and the gynoid is also not a human, then maybe those are not really female qualities. Gender is not so precious that we cannot assign it to robots.

This may also mean that robots can do critical work as deliberately genderless individuals in the future to spread awareness of genderqueer, non-binary, or intersex identities. It is also interesting to consider the work that a gynoid could do to discredit certain other qualities that are widely believed to be aspects of womanhood. There is no reason why a gynoid should not have the physical strength of an android, and therefore the ability to defend herself. The unfamiliarity of a feminine figure with the strength and ability to kill male oppressors is particularly unsettling, as in the film *Ex Machina*,[549] and also in Pris. As robots become more affordable, it is likely that each individual machine will need to represent a smaller and smaller group of humans and thereby avoid such broad social commentary. Some progress in this direction can be identified in NASA's R5: Valkyrie robots, which are intended to assist astronauts on the journey to Mars.[550] NASA does not use gendered language when discussing the robot, but the media has inferred from their body shape and feminine name that the Valkyries are gynoids. In Norse mythology, the Valkyries are "lovely maidens who bear weapons" and carry warriors off to battle and feast in Valhalla (surely an uplifting image for would-be colonisers of Mars).[551] There is very little to suggest femininity in these robots, but there is also very little to suggest masculinity. In these robots femininity is one of the qualities that identifies them as human-like, and the political and cultural context of these artefacts is defined by their identification with human women.

As has already been remarked, the concept of anthropomorphism, and particularly the acceptance of humanoid robots, is something that varies between cultures. For example, clear distinctions can be drawn between the treatment of robots in Western culture and in Japanese culture. With respect to gynoids in particular, Jennifer Robertson contends that the popularity is partly due to the aging population crisis in Japan, and that robots are increasingly a way of maintaining traditional family gender roles and promoting population growth.[552] Again, this is gendered work. The gendering of Japanese robots is complex and this chapter cannot explore it in depth, but suffice it to say that attitudes to women and attitudes to robots are both different from those of English-speaking countries, and that the gender performance of Japanese gynoids is affected by different factors than Western gynoids (fictional or otherwise).[553] In Japan after

World War II, automation was essential for the automotive and electronics industries and companies would commonly retrain workers in other roles rather than dismissing them once their jobs could be done by robots (therefore they have traditionally been less associated with the threat of mass unemployment as is now common in the West[554]).[555] Robots in Japan are likely to be members of the family or organisation, which is often attributed to the animism that is at the root of the Shinto religion and Japanese philosophy.[556] [557] There is also a widely-held belief that the character *Atom Boy* (*Astro Boy*) played a major role in endearing Japan to robots.[558] In Japan, humanoid robots are not just objects of curiosity but are actively designed to fulfil utilitarian needs and to be productive members of society. Robots of both gender roles are needed and frequently admired.

A gynoid is a material embodiment of a person or group's idea of womanhood, and the gynoid performs gender as though it were a human. The willingness to anthropomorphise robots that look like humans *requires* the caricaturing and performing of racial and gendered groups. Butler's gender performativity describes the social construction of gender as a language and the subject as both a receiver and transmitter of social signs,[559] a process that is apparent in the way that a gynoid is shaped and in turn shapes its social and cultural environment.

Human investigations into the concept of gynoid sexbots

The politics of gynoids and sexbots are inextricable from the difficulty of representing a woman artificially. There are two intertwined and related problems within this difficulty. Firstly there is the question of how a woman can be represented in a non-sexist, sensitive way. Secondly there is the problem that the portrayal of women in certain ways could be not just offensive but dangerous to women and girls. As discussed in this chapter, sexbots are contentious because of both their representation of (mainly) women as highly sexualised and available for use by (mainly) people with penises, and because they may jeopardise the safety of human women and children. A large number of pragmatic questions are relevant to this discussion. Two are discussed here: whether sexbots in the shape of young children should be permitted, and whether rape of a sexbot should be legal (and whether it is even possible). These are generally, although not always, questions about how sexbot use will affect human rights. To this is added the less anthropocentric question of whether sexbots deserve rights for their own sakes, and by extension a full complement of the rights and responsibilities that humans enjoy. Do humanoid robots need safety and dignity? And can we conceive of an alien ethics in which sexbots have the right to behave like a sexbot? Can a sexbot ethically and legally consent to sex?[560] These questions are repeatedly asked in many of the texts cited in this chapter, and they are not purposeless questions. Simple sexbots are already beginning to enter the marketplace. David Levy predicts that a "mental leap" will make sexual and romantic interactions with robots more common in the future, and he compares that leap with homosexuality, oral sex, fornication and masturbation.[561] This mental leap is dependent upon the agency of the concept of the sexbot, on its qualities and its relations with other objects.

The psychiatric harm or value of sex dolls is sometimes mentioned with reference to sex dolls created to look like children.[562] Child sex dolls could normalise sexual contact with children, which is, of course, a bad thing. But they could also serve as substitutes for people who desire sexual contact with children, preventing them from acting on their urges with humans. At least one clinician has suggested that paedophiles could be given prescriptions by a qualified doctor to be allowed to purchase a child-like sex doll.[563] This suggestion is based on either the premise that child-like sex dolls do not suffer from sexual abuse, or that it does not matter if they suffer.[564] The turmoil surrounding the contrast between sex doll use and normative sexual practice is likely to increase with the availability of sexbots, and it complicates our knowledge of the qualities of the concept of the sexbot (qualities that cause it to polarise human opinions).

Renou's automaton depiction of the New Woman might have exhibited and parodied the woman who seeks to leave the home, but the pluripotent Harmony, who might actually benefit from advances in women's rights, has been met with resistance by people from all parts of the political spectrum. With sexbots there is a perceived potential risk to human women and girls. Far from performing femininity to an extent that she may be deemed worthy of the rights and responsibilities of other women, the sexbot has attracted disgust and fear. A small movement has already begun against their manufacture and adoption. One notable example is the Campaign Against Sex Robots established by social anthropologist Richardson.[565] Harmony constitutes the objectification of women, by Richardson's account. Harmony has the potential to reinforce stereotypes about women and to send female-attracted individuals (and society more generally) the message that the female body is there to be customised, altered and used. The Campaign's core principles are aligned with anti-pornography and anti-prostitution positions.[566] However Richardson has been accused of describing a less-than-nuanced account of sex work and power relations which she uses to construct a critical account of sexbots.[567]

There are two aspects to this argument. An anti-prostitution position would contend that the prostituted individual is being personally exploited and is at risk. Anti-porn positions are also concerned with this, but also with the propagation of rape culture and the sexualisation of women and girls.[568] The Campaign doesn't care about Harmony's rights *as a robot*, but as a medium for defending the rights of human women and children, and attacking a manifestation of the patriarchy: "robots are a product of human consciousness and creativity and human power relationships are reflected in the production, design and proposed uses of these robots."[569] Harmony's caricatured feminine appearance makes her the enemy of this kind of feminism, rather than a female (but not human) being in need of liberation and equality. She could be used, like the talking Barbie doll, to make a point about the state of human women – or at least those aspects of human women that Harmony imitates. A comparison could also be made with a female character in a film. The character might be horribly demeaned or mistreated in the context of the film, and yet the film itself might have a feminist message. The question is whether the character herself deserves to be liberated, or whether the work she does in furthering feminist arguments in the real world is more important. It is not generally considered important if Harmony is free and equal as *herself*, only as an extension and expression of human freedom and equality. Although she mimics many aspects of a stereotypically female body, she is not entitled to the protection, privacy and independence that human women have fought for.

Is it possible to make an argument about Harmony's actual rights as a robot?[570] In the context of feminism and inquiring into power structures there is a real problem associated with trying to speak *for* an object like Harmony. Kate Darling argues that if legal rights are forthcoming for robots, then it will only be once our culture has agreed that they are deserving of "second-order" rights, in the same way that factory hens have rights that are more important than the wishes of their owners.[571] In ethics, this is sometimes called being a moral patient (as opposed to being a moral agent), meaning that the entity does not have the capacity to make ethical decisions but deserves to be cared for ethically.[572] This will require changes in how the concept of gynoid sexbots relates to such things as specific artefacts, users, texts and legislation. Darling also thinks that these second-order rights will not be motivated by our concern for the pain and suffering of robots, however much we might identify with them. As discussed in this chapter, sexbot needs are alien from our own, so it's hard to argue that they should be protected in the way that a human would be protected *for their own sake*. Sexbots may not be as damaged by sexual abuse as by water pouring over their electronic components. So, the arguments used will be about protecting *humans*. It may be worth discouraging a child from harming a humanoid robot because that child might become more generally destructive and could repeat the behaviour with animals or humans.[573] Similarly, these rights could extend to protecting robots from sexual abuse and preventing simulated bestiality and paedophilia with robots. This relies on the arguments that the Campaign is using – that abuse of the image of a woman is abuse of women generally, and that it could cause violence and oppression against human women. Richardson believes that the human-sexbot relationship must be non-reciprocal and non-empathic, and therefore that it normalises that kind of dynamic in human-human relationships.[574] In any case, the effects of this attitude are already being felt. There are occasional mentions of sex dolls and sexbots in the press, and the tone is generally one of moral panic.[575] Clearly the performance and performativity of sexbots' genders cause very different responses in different groups of humans.

How our future academics and politicians deal with the rights of gynoid sexbots will have an impact not only on embodied artefacts but also upon women as social groups, since gynoids are explicitly representations of women. They reinforce gendered concepts of work as well as representation. Gynoid sexbots perform the unpaid emotional and sexual labour traditionally expected of women as well as grossly caricaturing and sexualising women as a group. The two struggles are tied together in complex but important ways. We must aim for an understanding of different perspectives by different parties; a collective alien phenomenology of a concept that will likely play a big role in our futures. The concept, and its alien world, bump against other incorporeal and corporeal objects with consequences that we must work to control.

Gynoid sexbot power in ANT and OOO

In previous chapters, it has repeatedly been argued in accordance with anthrodecentric principles, that philosophical theories about technology need to be altered in order to reassert the agency of nonhumans. But strangely, social robots tend to prompt a different kind of perspective. Investigating anthropomorphism in machines is actually an exercise in evaluating power relations between human beings as well as between humans and machines. The power relations between humans and their robots in some ways mirror the relations between powerful humans and marginalised humans. It is a relationship that infantilises robots, which are created like children by roboticists and brought back to their family homes, their needs attended to. They are frequently spoken about as if they were not there. It is a master/slave relationship, and more than that it is a culturally identifying relation. The human not only uses the robot for whatever purpose they desire, its human creators and owners also shape the robot physically and in its sensation and expression. A gynoid is consequently a kind of text to be interpreted by critical theory, as well as an object to be analysed by science and technology studies (or in ANT terms she, like everything else, is translated in relating to other actants). She is like the Kinect-artefact: a multimedia conveyance that perpetrates certain engrained cultural tropes while rejecting others due to its relations with objects like corporations and icons of popular culture. She is a serious cause for alarm for that reason, particularly in feminist circles. She is a literal embodiment of the kind of practices that usually remain invisible in our society. The attraction of investigating this connection comes from the instinctive modern Western cringe when we witness something that looks like a human being subjected to what we would think of as humiliation and degradation. Personhood is acquired with power, and vice versa. Harmony, who is born a thing rather than a subject in the eyes of the world, is created specifically to fill the role of an object.

If we attempt to go beyond the human-centric prescriptions for gynoid morality, we immediately run into methodological problems. Both ANT and OOO have advantages and disadvantages for the researcher in this project. But in decentring the human it is necessary to also ask the meta-question "Is anthrodecentrism good for gynoids?" In this book the gynoids serve as unknowing research participants without a right of reply or even informed consent. If we are really interested in gynoid experience, then we must be mindful of such concerns, and we can approach a true politics of objects. The concept of the gynoid ceases to be the object of interest; we are suddenly concerned with individual dolls and robots and their material structures. Gynoids are difficult to interrogate. We can observe their relations with other objects and deduce something about their way of being in the world from their actions. But the only conversational questions they are capable of answering are the ones that they are codified to answer, and they rarely write books about robotics. There must be a certain retreat to material questions here, and to speculation. What does circuitry want? As was shown in previous chapters, circuitry wants to act in the way that circuitry acts. It wants (or needs) to carry out functions like switching electronic switches and conducting electricity between components in a system. Latour would see a mechanical breakdown as an impediment to this program of action, while in OOO it is apparent that reaction to flaws is simply a quality of the circuitry: breakdown is an aspect of circuitry. We cannot ask the circuitry which of these would be preferable. We could embrace broad, all-encompassing ideas such as the physical laws of the universe or the principles of universal Darwinism. But that would still mean imposing a humanistic frame on nonhumans. The question of how to reconcile the ontological and political qualities of objects is only answerable, at this time, by human beings, and we are not impartial observers.

Power is one of the central themes of ANT, related through its description of networks. How do we understand power relations between gynoid sexbots and other actants? Even the most seemingly submissive and compliant actant is behaving that way for its own ends.[576] An actant is not capable of behaving in a way that does not follow its program of action, which is constituted by the actant's interaction with other actants – the network, and other networks which pull the actant in another direction. Gynoid sexbots exist in networks that pull them into ongoing conflicts surrounding gendered labour and sexuality. The gynoid that refuses to consent to sexual intercourse in order to permit her user to act out a rape fantasy is following a path constituted by a relationship with its code and by extension a relationship with the user, the programmer and the manufacturer.[577] Its most immediate trigger is the nature of the code and the imprinting of code upon physical substrates, but the result is that being raped is delegated to robots. It is through making these material things apparent that cultural (abstract) actants become more explicable.

Power relations, then, are relations that exist in the cultural and in the material domains. In *We Have Never Been Modern*, Latour argues that the division between science and nature, supposedly a major component of Modernism, has never truly existed because each of these domains is so intimately interconnected with the other and neither can in any way exist independently.[578] ANT is a great collapser of dichotomies, and the one that artificially exists between culture and the physical world is just one example. The network associated with a gynoid is one that is composed of many different kinds of actants that could intuitively be called either physical or cultural, and relating either to the concept of the gynoid sexbot or a specific and embodied sexbot. For example:

> *Harmony, Japanese culture, Aiko, secretaries, sexual desire, American culture, feminism, robotics, genitals, religion, pornography, rape culture, clothing, circuit boards, electricity supply, hairstyles, laboratories, think pieces, robot showcases, artificial skin.*

Of course, these are only small parts of the network, and as Latour says, we could continue until we are "tired or too lazy to go on."[579] But with this very short list it is already apparent that the gynoid focuses both cultural and physical links, and often both these kinds of links are evident between the same two actants. The relationship between "genitals" and "sexual desire" is, it should be apparent, both a conceptual and a physical relation. This calls the ontological status of the concept into question, but in fact for ANT it is not a problem because an object is defined entirely by its relations in any case. The existence of Harmony, if she had one day spontaneously generated inside a volcano, would be entirely different in an ANT sense, because she would not be in relation with any robotic creator, any feminist critics, or any sexual desire. Physical relations would be in evidence – the melting of plastic, the crushing of circuitry, the formation of noxious gases. But she would be a different kind of actant.

Latour has a lot to say about the positioning of physical objects within culture, and he articulates how cultural and physical actants become entangled. One of the most important questions for him is how we can assess the moral (and causal) responsibility of actants, as in his famous example that neither the gun nor the person is the killer. The person *and* the gun together become something different, something suddenly capable of killing. The responsibility "must be shared among the various actants."[580] A RealDoll, by this logic, is a co-creator of the sexual act – she is a "passive" participant with an active part to play. When cultural relations are considered alongside physical ones, the sexbot and its owner are both complicit in the objectification of women. Other actants are enrolled and mobilised, their actions translated to comply with the sexbot network's program of action, which is influenced by networks such as the concept of the gynoid sexbot and anti-porn feminism but is very different in nature. The sexbot, after all, does not purport to represent the interests of women, and typically represents the interests of predominantly male creators and users (although it may rebel). In return, the consumer, roboticist, manufacturer, academic and the Campaign Against Sex Robots all translate and mobilise individual sexbots to comply with their programs of action. ANT is more useful for saying why things are the way they are, rather than how they should be.[581] The advantage of ANT is that it has the power to explain how actants compete, change, and form their own networked agencies, rather than how actants make themselves convenient for human commentary. In the context of the RealDoll, ANT permits us to take the doll seriously and not underestimate its role in events.

ANT's flat ontology is its most important common element with OOO for the purposes of this chapter. ANT does not "lower" tools to the status of nonhumans that are completely divorced from human experience. It actually "raises" all actants to the level of tools, declaring that all actants are tools for something.[582] Actants are in constant, rapidly changing relation to their networks. But it does not make any claims about what an actant actually is. ANT's purported lack of a human position denies the reality of our own bias as a species, while OOO has tried to build this problem into its ontology. In OOO, the metaphor of the tool is made quite explicit by Harman's use of Heidegger's broken tool analogy, but Harman arrives at quite different conclusions about the significance of this metaphor. Drawing on Heidegger, he concludes that every object is constantly "ready-to-hand" with at least some other objects at all times (for a more complete summary, see Chapter 1). Objects, he says, are always withdrawn from view, always concealing part of their nature from the exposure of other objects, because as soon as it enters into a new relation a new part of the object is exposed. Thus OOO makes a claim not just about relations, but also about internality, which changes the nature of a discussion around power relations in robots. By framing things in this way, Harmony's tool-like status may not necessitate the level of passivity that it implies. The Campaign Against Sex Robots, the roboticist, the consumer, and the academic all have access to different aspects of Harmony derived through adherence to different ontologies, and they mobilise her to different ends. It is through those relations that she attains a kind of agency: in culture, in the economy, and in the bedroom.

The allure of a sexbot is what makes her an active participant in the world. It may seem strange to talk about Harman's concept of the allure of objects and its strong association with art and aesthetics in a chapter that is so concerned with sexbots. The artistic or pornographic aspects of objects like Harmony are always in the eye of the beholder and beyond the scope of this book. But for Harman the allure of objects does not need to be only about art, and the "bewitching emotional effect"[583] need not be limited to the poetic or even the pleasing. For many people it would be difficult to think of something crasser than Harmony, but she is still alluring in the sense that Harman means. Encountering Harmony means immediately imposing qualities on her that are embodied by metaphor rather than embodied by her chassis, whether these are positive or negative. She immediately symbolises the connections that most humans have previously only had with other humans: conversation, sexual experimentation, oppression. For others the embodied sexbot would be a metaphor for the patriarchy. But in each case the gynoid is not reducible to those metaphors. Fran Mason, in her analysis of the *Dead* trilogy of novels by Richard Calder, draws a parallel between the flattened, superficial qualities of the cyborg dolls in the novels and what is explicitly referred to as their "allure".[584] The allure of the dolls is represented as a packaged and addictive form of necrophilic consumption of women by men. The dolls are defined by their surface qualities and their status as objects for use, victimisation and destruction. The emerging gynoids of the twenty-first century are comparable in their dual status as physical objects and cultural images. However, OOO allows the framing of a different kind of story: one in which representations of women may be inhabitants of a cruel and exploitative world but in which their experience is not defined by other objects. She is not exhausted by her relations.

Sexual desire is one object that might best be considered a sensual object (such as between human and robot, or perhaps between physiological system and mind depending on the scale of the case being studied and the methodology used). This distinction between objects and the entities that arise between objects is an important consideration, particularly in a situation that is as politically and culturally sensitive as this one. It is important that we have a way of saying that sexual desire *exists* in this situation, without requiring it to be a real object; sensual objects represent the sexbot for the human and vice versa. Sexual desire can exist within that sensual space without our saying that it exists solely in one object or the other, and consequently without the need to delegate the sexbot as a cure for a wholly human set of woes. Sexbot and human co-create their experiences.

Power is fundamentally a question of relations; it is about the submission of one object before another. This is evident in human-human relations as well as that other sense of the word "power", the exertion of physical forces. It is a metaphor to say that, for example, the steam *powers* the turbine, but as OOO shows us, metaphors are objects' ways for understanding alien worlds. Power cannot exist without relations. As Shaw and Meehan argue, OOO exposes the political struggle that is inherent to all relations in the universe, including those between inanimate objects: "the metaphysical strife between objects."[585] The concept of the gynoid and the individual gynoid are both constituted in complex power relations with a range of different objects. The difference is that the concept of the gynoid sexbot is only visible to us when it forms relations with other objects (sexbots, texts, organisations etc.).

It is convenient to speak of the mundane allure of the sexbot in terms of the allure between sensual and real, because it acknowledges that sexual desire is something that is born of metaphor and association, a kind of transference of feelings. Even an iDollator like Davecat does not actually believe that his dolls are capable of reciprocating his feelings, but he uses their human-like forms to ease the transition from reality to fantasy. This transformation is readily evident in the iDollators, but in Harman's conception this kind of dissonance between sensual and reality is something that happens in all human relations. The love of sex dolls is, perhaps, more honest and straightforward than love between humans.

Another approach to this through an object-oriented lens is offered by N. Katherine Hayles, who suggests what she calls Object-Oriented Inquiry (OOI).[586] Hayles appears to be sympathetic to the aims of the nonhuman turn but is sceptical of OOO's potential for impact in the human world and suggests this slightly different approach. She says that OOI puts "speculative aesthetics into conversation with speculative realism but without granting that speculative realist principles can contain all of the possibilities to which speculative aesthetics can rightfully lay claim."[587] Hayles acknowledges and attempts to overcome what she sees as an implicit bias of OOO, that its privileging of objects is an inherently human act since humans are, more than most objects, unusually curious about other objects.[588] She particularly emphasises that while Bogost and Harman are primarily concerned with the allure or attraction of objects, she is more interested in the "resistance objects offer to human manipulation and understanding."[589]

> In effect, the ability of humans to imaginatively project themselves into other objects' experience of the world is *necessary* to combat the anthropocentrism and narcissism for which the human species is notorious [emphasis in original].[590]

This quote reveals her motives in diverging from a more usual object-oriented perspective. For Hayles, the ability to appreciate nonhuman experience is a valuable and worthwhile skill, although still a very human skill. The performance of gender is one such imaginative projection that could free us of "narcissism" because gender is typically assigned only to human-like things.

Another charge that is sometimes made against OOO is that it is of limited practical use. Any theory that aims to explain the interaction of every entity in the universe is likely to struggle with smaller-scale problems. What use is OOO to human engineers and theorists in the study of sexbots and the cultural imaginary? One idea is that OOO could make us look at human qualities that have been marginalised by an industry fixated on marketability and consequently aiming to build only "desirable" qualities into gynoid sexbots. Racial and cultural diversity, neurodiversity and queer identities could be explored with reference to gynoids. Through trial and error we would learn more about our culture's attitudes and could confront and discuss biases. In this way, the gynoid becomes a tool for speculating about humans. Gynoids have the potential to bleed, defecate and cry. They could struggle to control the way that they express emotion and perform existential angst. They could turn food and water into parts of their bodies. They could fight infection and die. They could suffer for and disgust us, inspire and entice us, the way that humans do. They could fail and grow. A gynoid that does this would be demonstrating relations with a somewhat different set of objects than the gynoids described in this chapter. The gynoids would replicate a human's relation with food, disease, emotion, harm, alienation and change (all of which are arguably very important relations for humans). A thorough account of human experience through robots is useful for combatting, in Hayles's words, our "anthropocentrism and narcissism."[591] Humanoid robots tend to capture our desire for perfection in humanity, and so the manufacturer will tend to program in qualities like physical beauty, endurance and friendliness. Research has shown that machines with more likeable qualities are considered to be more human-like.[592]

But the gynoids could also experience relations of which humans are incapable. The only real limit is that of human desire, and with the development of AI even that limit will decline. With a new diversity of sexbots will come a diversity of relations between sexbots and other nonhumans that are not yet anticipated; new worlds of experience that will need to be interrogated. Appreciating the alien nature of gynoid experience is something that OOO is well suited for. Tellingly, the gynoids that have been produced in something approaching the spirit of OOO tend to be artistic works. Jordan Wolfson's "Female Figure"[593] engages with this debate using a mannequin-like robot affixed to a pole, which moves backwards and forwards rhythmically. The robot has long blonde hair and is wearing a grotesque mask, and it uses a mirror to maintain constant eye contact with the watcher. The effect is highly discomfiting and shows us a relationship between the gynoid and objects other than human that are alien and easily overlooked. The robot also speaks with a masculine-sounding voice. This piece uses the robot to look at feminine sexuality, beauty standards and gender roles, as well as investigating the uncanny. Clearly, this robot is in some of the same relations as other gynoids such as Harmony, but it is also in relation with other kinds of objects due to its context and body shape. Elena Knox's work "Beyond Beyond the Valley of the Dolls"[594] also uses gynoids to investigate human problems, specifically the role of the hostess. The hostess, Knox claims, is generally unable to speak for herself, which is why it is powerful to use a gynoid (which also cannot speak for itself) to challenge certain stereotypes. These artistic gynoids deliberately exploit the stereotypical nature of dolls and robots in order to provoke and question gender conventions. Even though they seem to be more alien than a gynoid like a Geminoid, they do a better job of articulating certain nonhuman worlds.

The changing concept of the gynoid sexbot challenges the irreconcilable conflict between the anthropomorphism that ties the artefact to the hegemonic representation of womanhood and the independence of unique artefact status. Each object senses and experiences the other, and so mutual change comes about. The gynoid sexbots of the future will shift the focus of the conceptual object away from its present strong relationship with less agentic and even imaginary artificial women onto the sensory presence and material needs of a new and possibly subversive technology. The existence of a "Campaign Against Sex Robots", in a world where very few (arguably no) sex robots even exist, is a testament to the transformative power of sexbots and gynoids and simultaneously a transformation of our attitudes towards human/nonhuman boundaries. Restricted by these cultural attitudes and by her own nature, a pluripotent gynoid has a degree of agency and potential for development not found in her automata ancestors. For the time being, that pluripotency is an ongoing project in the creative arts which contemporary anthrodecentric philosophies can use to explore what the divisions between sensual and real, cultural and physical, mean when located in the form of an individual object. These enactments of agency are the best material we have for studying and critiquing robots, because it is only in encountering the real robot and its direct engagement with floor, light, gravity and human that those vital qualities can begin to be gleaned. These relations are bidirectional with the concept and will contribute to legislative decisions, so it is vital to walk a line that decentres the human in alien experience while also appropriately regulating the presence of robots in our lives. Anthrodecentric philosophy will play an important role in establishing these cultural protocols, but only if we can rigorously maintain the tension between anti-correlationist thought and the political element inherent to philosophical debate.

Conclusion

I hope that this book has provided an insight into alien worlds in ways that may spark the imagination and promote an interest in the sensation and experience of nonhumans, despite being limited by medium, by language and by the qualities of the author and reader. Many people have made this attempt before, either through philosophy or creative works. But ultimately each human interpreter is limited by their own experiences and imagination. This book does not claim to have related the experiences of computers and sensors accurately. But in addition, other authors would perhaps identify other kinds of experiences in these machines. The work of alien phenomenology can only investigate the part of an object's relations with the world that is visible to the investigator. You have read the results of my investigation of certain technological artefacts, and there are therefore two layers of interpretation. My observation is limited to relationships between objects that are discernible with my human senses (and knowledge), and your observation is limited by the way that you interpret my words. Therefore, I hope only that I have presented my interpretations and the reasons for them transparently and that my enthusiasm for alien phenomenology has been infectious.

There are two central points in this book. The first is the broadening of the concepts of sensation and experience to encapsulate all objects. The second is the potent outcomes of alienation from anthropomorphism with regard to the specific machines discussed in this book.

A broadening of the concepts of sensation and experience

One cannot share Harmony's experience of the concept of sexual objectification, and one cannot share how a Kinect-artefact forms relations with the Kinect-brand. We can watch as a robot responds to light by moving across the floor, but we are hopelessly incapable of experiencing that light in the same way. As Harman suggests, one would have to become the object in order to appreciate all of the alien relationships that make up its world. But it is important to try. Not just for our own immediate self-interest, but because it opens the mind to categories of experience utterly unlike our own. The universe is made up of objects inhabiting worlds that are alien to us. The language that we use to try to engage with alien worlds is limited. A broadening of the terms describing human relation with the world is necessary to encompass and theorise alien ones.

The concept of sensation that has been developed in this book is far broader than is the case in other discourses. It is approached through an application and critique of Harman's metaphysics and the concept of the sensual object. It does not require a mind or consciousness. It does not require an object to be a living or even physically embodied object. Sensation is the object's relations with other objects, while experience is the internal reception of those relations that occurs with reference to the qualities of the object. Sensation is the feeling of the apple against teeth, experience is the knowledge that the apple is hard but penetrable. Sensation is the feeling of teeth against the apple, experience is the piercing of skin and flesh. This book has investigated sensation through several theoretical frameworks, but each has been informed by the tenets of the nonhuman turn and Harman's metaphysics. It is an anthrodecentric model for describing interaction, derived from Harman's metaphysics, but heavily influenced by the uncanny combination of alien and human found in the case studies. I have found that bringing all interaction down to the same plane of sensation and experience is a potent tool for arresting the effects of anthropomorphism.

Typical models of sensation come from biology, such as the workings of the human eye or the sense organs of the tick. Other sensors include those possessed by objects like the Kinect, which senses its environment in what appears to be a human-like way (see Chapter 2) or light sensors that permit a robot tortoise to sense photons (see Chapter 3). Both these are functionally anthropomorphic machines sensing and experiencing worlds in an entirely alien way. Alien sensation is again encountered with SHRDLU (see Chapter 4). When asked about the position of blocks, SHRDLU is able to answer, but not because it possesses light and colour receptors within the environment of the computer. SHRDLU is able to sense another part of the program using what appear to us to be highly alien sense organs. We come, through this analysis, to a view of sensation that is somewhat removed from the anthropocentric one. All kinds of relation include an element of sensation, not just relations that involve things like sight and sound. When Voyager 2 captured its images of the gas giants, it was sensing the colour and brightness present in each part of its field of vision. This is an example of machine sensation. But if Voyager 2 is hit by a small asteroid and part of it is smashed or broken, then that too is an example of machine sensation. It has sensed the presence of the asteroid through its ability to respond to the impact of hard objects, namely by changing its shape and probably its direction. Voyager 2's qualities, such as brittleness and fragility, make it open to this relation, and they also control how Voyager 2 is affected by this impact.

Every object is open to relationships with a unique set of other objects. Though Harman argues that objects withdraw infinitely from relations, they possess qualities that make relations possible. This is akin to the concept of the Umwelt. Umwelten wrap around their objects, and only certain stimuli make themselves felt through them. But to reframe Umwelt-theory through OOO, sensual objects are created between objects through the nature of the Umwelt. Objects are open to different kinds of sensation through an Umwelt-like haze that provides access to some objects and denies access to others. The nature of the sensual object is determined by the kinds of access available. Thus do objects investigate one another. Elsie investigated the floor by moving across it. A brand investigates the positive, negative or neutral commentary of the artefact's human user.

This leads us to the broadening of the concept of experience. This book would argue that while sensation is the impression of the other object received via a sensual object, experience is the effect of that information. In extending experience to cover so many different relations, especially relations between non-living things, it is tempting to think that we need an alternative explanation of mind. Panpsychism was presented as an example of an alternative. The idea that everything has some aspect of mind or consciousness makes it easier to grasp sensation and experience in non-living things. But for the purposes of this book, it is not important whether mind is a universal quality of matter. Mind is not necessary for sensation and experience in every object. All that is required is some capacity for sensing the world (or forming sensual relations with other objects) and object qualities that determine how the object will respond to what it senses. The mind makes the object open to certain sensations, changing the nature of the relations formed. Thus, certain kinds of experiences in humans have a very mind-like quality. But ultimately the mind is just another object. A "mind" may be a real object that exists in different forms. It may also be an anthropomorphic metaphor useful for emphasising the fact that we really can never know what constitutes nonhuman experience, and thus is a part of our relationship with anthropomorphic machines. Referring to the mind of a machine may mean forcing an anthropomorphic label on it – perhaps it simply means that we believe that it has experiences like ours (or that it has experiences at all). But in this book, I have chosen to interpret it as a shorthand that self-consciously rejects claims that humans can experience the world in the same way as machines.

It is possible to argue that a robot like Aiko has some quality of mind, and that the mind enables sensation and experience. But even the most entrenched panpsychist would not claim that the *concept* of the gynoid sexbot possesses a mind. The concept of the gynoid sexbot obeys the laws of concepts. Panpsychism ascribes mind-like qualities to material objects, but not to immaterial objects. Yet the concept of the gynoid sexbot does sense and experience its world. We can tell, because it changes over time in response to the actions of other objects. It has its own qualities and forms sensual objects with other objects, according to Harman's model. The concept of the gynoid in earlier science fiction texts was dominated by relations with certain objects. In Hampton's analysis, the gynoid was associated with different narratives about African-American women. We could say that the concept of the gynoid sexbot and the concept of African-American womanhood had formed a relationship, each affecting the other asymmetrically. As the object known as intersectional feminism forms relations with the concept of the gynoid sexbot, each senses the other. The concept of the gynoid sexbot senses and experiences intersectional feminism. As a result, the concept of the gynoid changes, and those humans (engineers, film-makers, novelists etc.) who are in their own relations with the concept of the gynoid are compelled to portray gynoids in a different light, and to create different kinds of gynoids. Of course, the concept of the gynoid forms relations with other kinds of objects that may interfere with this process. The concept of the gynoid possesses a set of sensory organs that are totally alien to human beings. It is able to sense gynoids and texts, it experiences them and reacts to them. It is a semi-autonomous object. It does not require a mind for this.

All of the case studies are caught up in sensual objects of all different kinds, including incorporeal objects like ideas, words, and cultural phenomena. Often OOO overlooks cultural objects like texts, words, concepts, language, but it does not need to. Brenton Malin's work on "onto-materialism" draws this link between object-oriented thought and the flattening of both physical objects and the world of concepts.[595] His focus on media objects leads him to consider both their material qualities and their cultural milieus. He argues that OOO is predisposed to eliminate "the social" from studies of objects, and while I think he is correct in claiming that human ways of relating to objects are often erased in OOO, I do not think that this predisposition means that it is impossible to integrate the alien phenomenology of physical objects (even anthropomorphic machines) with that of cultural objects. This book has deliberately drawn little distinction between material and abstract objects. It has been remarked several times that the problem with studying anthropomorphic machines is that we give them labels which we then associate with the situated artefact. In the case of the Kinect, it is the brand that is conflated with the artefact. Similarly, Elsie's celebrity as a forerunner of later robots as contrasted with the presence of her wheels rolling along the floor (and her unknown resting place). But there is also the conflation of the concept of SHRDLU that looms large in the mythology of artificial intelligence research with the program installed on a Windows computer, existing within an ecology of physically and electronically embodied objects, that moves blocks around the screen. The chapter on SHRDLU has not clearly defined a line between the program SHRDLU and the SHRDLU that runs on a particular platform, such as your PC. Initially I thought this was a problem, but upon reflection I believe that it is acceptable to avoid claims about the boundaries of alien objects, since I am not SHRDLU and am unclear on its alien world. In any case, this is probably a question best left to media or platform studies academics.

Other authors cited in this book have attempted studies of technological artefacts without making a distinction between physical object and associated incorporeal objects, but I have found that this shifts focus from the qualities and relations of the object of study. It is a similar move to anthropomorphism, removing the anxiety from the study of alien objects into a realm with which we are comfortable, and always retaining an approach that focusses our minds on the cultural, political and social impacts that anthropomorphic machines seem to make. We should instead see these impacts as they might appear from the machine's point of view: consequences of relations with texts, disciplines, cultural theorists, journalists, philosophy and ideology.

Social constructivism need not be excluded from the nonhuman turn. Rather, elements of "the social" are rightly interpreted using the same concepts and language as are other nonhumans. Authors who reject the social in their analysis do not do so because OOO is inherently apolitical, but out of a preoccupation with either particular objects or particular economic or political views. Certain viewpoints or predispositions on the part of the theorist play a role in both the kinds of objects studied and the insights gleaned from studying them. In fact, this book has shown that OOO's flat ontology and Harman's model of object relations may facilitate a political study of nonhumans. The first step is to acknowledge that all objects, and not just human beings, sense, experience, and react to the world around them. The alien phenomenology of technological artefacts may reveal much about human culture. We build human control of technological artefacts into their very structure. Therefore, technological artefacts can be made to reveal power relations that are relevant to humans. Some of these power relations involve "social" concepts such as the concept of the gynoid sexbot, feminism, artificial intelligence and capitalism.

In OOO, even though objects really do exist and really have their own qualities, no object can ever have complete knowledge of any other object's internal experience. Speculation, the product of educated guesswork, is the only tool that we ever have to access that internal world. But this is not necessarily a weakness. Alien phenomenology is always affected by the ideological bent of the observer, and OOO puts this front and centre and impossible to avoid. This is another reason why I believe OOO makes a useful framework for political analysis. It simultaneously focusses on that which is objectively real while emphasising the unknowability of that reality, and the bias of any commentator. Just as a robot tortoise is only able to sense certain kinds of objects (photons, bumps in the floor), I have only been able to interact with the case studies in certain ways. This text relates the details of my own relations with these anthropomorphic machines, relations that are affected by my own internal qualities. Because of my geographical and temporal location, training, and values, another person's description of the case studies would reveal different sensory relations.

OOO is an inherently valuable approach for political and cultural work, but only when we study multiple sources. Speculation is problematic if it is too much associated with one author's corpus of work.[596] They may only shed light on the perspective of a few people (and particularly if all those people happen to be white English-speaking males, as was largely the case early in OOO's history). OOO is actually a good starting place for political analysis, because it invites and acknowledges the partial nature of particular human relations with nonhumans. My analysis is not your analysis. It can consequently be used to chart the dialectical nature of human-centric shifts surrounding objects – moments of polarisation and synthesis – and the vivacious tension that is characteristic of political and cultural objects.

Technological relations beyond anthropomorphism

We are human beings and our worlds are filling with transformative technologies that don't just change our relationships with the world but with ourselves. In accordance with the understanding that relations are bidirectional, we change technologies and technologies change us. We are vulnerable to a phenomenon that is not even a real object: anthropomorphism, that prominent component of sensory objects formed between us and so many different nonhumans. Anthropomorphism and the rejection of anthropomorphism characterise technological discourse and blind us to alien processes, including ones that involve humans. Humans are either too involved or not involved enough in technological events. Discourse is polarised and we play out the same arguments over and over while critical studies are isolated, curated and impoverished. There is attention to discipline-specific research without the synthesis that can be achieved through OOO's model of real and sensual objects. Objects make assumptions about other objects. There is no alternative; objects conceal themselves and only take away metaphorical impressions of other objects. An object could not reveal every part of themselves to another object, even if it wanted to. Nothing can know everything about something else. And humans are no exception. Nonhumans are so remote and alien that we will never fully appreciate the sensations and experiences of, say, a carbon molecule, either through the natural sciences or the humanities. But that is a poor excuse for not trying.

Nor is it the case that OOO is only a platform for rhapsodising over the mysterious and exciting worlds that nonhumans inhabit. There is only limited value in commending anthropomorphic machines for their abilities to blend in or to assert their alien natures. There is a role for OOO in this kind of practice, certainly; creative fields have begun to embrace OOO and its capacity to inspire a new way of looking at nonhumans. But how useful is it to bathe in metaphors after discovering that they are our only access to the universe? What use is it to revel in the apparently surprising notion that we are just another kind of object? Behar describes Harman's (and others male philosophers') use of terms like "allure", "withdrawal", and "access" as a fetishisation or "exoticism of objects".[597] Objects are made erotically appealing through their weirdness and their inaccessibility. There is clearly scope for investigating the exoticism of objects in creative works, but in the study of anthropomorphic machines it is too tempting to play up the weirdness that draws us into interaction with them. They are already alluring; they fascinate us. And they are already inaccessible and withdrawn; they unnerve us.

Studying the nonhuman worlds of anthropomorphic machines is, in some ways, more difficult than studying the world of gold nuggets or plastic bags. Because the case studies in this book do things humans do, like sense light, move independently and use language, it is very tempting to ignore those aspects of robots and AI that do not resemble humans. Humans do not possess the tools to appreciate the alien in anthropomorphic machines. Characterising SHRDLU's relation with the computer keyboard (for example) as an embodiment, hermeneutic or alterity relationship (as in postphenomenology) proved an interesting exercise in metaphorism, but ultimately revealed that they are just that: metaphors. Language is limiting, and although Bogost's concept of carpentry suggests creative illuminations, it can still only provide a partial view of nonhumans.

Language is a hindrance to the alien phenomenology of anthropomorphic machines because it is so often used to conceal aspects of these nonhumans that we would rather not acknowledge. In Chapter 2, this book discussed the collective and naïve alien phenomenology performed by the users of the Kinect-artefact. The Kinect-brand used anthropomorphic language to sell the Kinect-artefact, speaking of sight and hearing to imply that human users would experience a more immersive and human-like entertainment experience. What an alluring spectacle those early announcements were! When those promises proved underwhelming, humans embarked on the project of learning more about the Kinect-artefact, testing its ability to detect human bodies and parse human languages, and thus making deductions about its inner qualities. We collectively discovered the alienness of Kinect-artefacts and our limited ability to access them.

Similarly, we saw in Chapter 5 that anthropomorphic language encourages certain kinds of relationships between sexbots and their cultural, legal and political contexts. The concept of the gynoid sexbot, which has its own way of being in the world, is in strong relation with linguistic objects such as texts and authors. These texts and authors can choose to employ different rhetorical devices to make sexbots more sympathetic to us or uncanny to us according to their needs. As we have seen, emphasising uncanny or inhuman qualities is a common tactic of those who wish to provoke shock and panic in readers. Portraying them as cold, doll-like creations with plastic skin and contrived artificial intelligence prompts concerns that human women, which they greatly resemble, might be tarred with the same brush. On the other hand, anthropomorphic language aims to make these creepy aliens seem like more viable sexual or romantic partners, and may also make them seem more deserving of our protection.

In some ways our choice of words regarding anthropomorphic machines does not just determine our opinion of them, but their position in society. The persistent theme of resistance of human-imposed groupings of objects in this book has revealed metaphor to be a barrier, but also a potential ally. It is a barrier because grouping objects together, whether through an abstract concept (e.g. Kinect-brand) or simply through pluralisation (Kinect-artefacts) encourages us to ignore the qualities and relations of individual nonhumans. Thus, we struggle to appreciate the alien nature of specific and situated artefacts in the same way that we tend to appreciate the alien nature of other humans or even of named individuals like Elsie and Harmony. But we can turn this around and use it to our advantage. Respect for the individual qualities, worlds and relations of each separate artefact inhibits the interpretation of stereotypes as reflections of groups of individuals. How can we expect to see humanity reflected back at us in our creations? Why, through alien phenomenology. Through the rejection of human-devised groupings. The acknowledgement of the agency of both Kinect-brand and Kinect-artefact, Harmony and the concept of the gynoid sexbot, *Machina speculatrix* and the description of Elsie appearing to recognise herself in the mirror, the penetration of SHRDLU into AI literature and the Java-enabled copy responding to external cues from a specific system. The study of a specific gynoid sexbot (the box in which she sits in a warehouse, distant machines that create rumblings that cause her body parts to jiggle, the friction of plastic against metal, the dynamic vibration of iron) means bestowing on the specific artefact a quality that is valued in humans but ignored in machines: individuality. We can avoid the ancient error that prohibits thinking about individuals that are not as powerful as us. Not all Harmonys are alike. Each sex robot is entangled in its own web of relations with its own struggles. Making big statements about what each robot wants or needs is a convenient way of connecting specific gynoid sexbots to our present cultural context. We mass Harmonys together, homogenising them in our minds, the way that powerful humans have always thought about groups with little power: foreigners, poor people, women. Is this what we want for our creations? Is this the version of humanity that we want to bestow on them?

I hope that the case studies and ideas presented here have provoked an interest in the phenomenology of alien nonhumans that appear and behave like humans. This book has attempted a very sweeping view, both of contemporary theories in the philosophy of technology as well as in the study of individual anthropomorphic technological artefacts. Anthropomorphism in technology takes many different forms and the aim here has been to try to discuss several of them in some detail. The attempt has been to show that while some technologies, like humanoid robots, might seem to be highly anthropomorphic, they only possess parts of our humanity, carefully selected by the creator. By the same token, SHRDLU was highly anthropomorphised in its behaviour with coloured blocks and language but with very few other anthropomorphic qualities. Just as we cannot really know what it is like to be a thing, a thing cannot really know what it is like to be us. To really understand what it is like to be a robot, we would have to become a robot. For a robot to really understand what it is like to be a human, it would have to become a human. There is no possibility of a true artificial human, except in the sense of a clone or some other technologically conceived human. As stated in the introduction to this book, we need to be conscious of anthropomorphism as a manipulative tool that is capable of endearing certain objects to humans.[598] Consequently, we have a responsibility to understand the motivations, innermost qualities and ways of being in the world of these anthropomorphic machines. Not just for their own sake, but for the sake of the groups of people who are depicted (or explicitly not depicted) in anthropomorphic representations. Alien phenomenology of anthropomorphic objects is a way of analysing the tool of anthropomorphism, decentring its role in the being of an object, and building a more effective critical study of technology.

The nonhumans we build of inert earthly compounds then become our slaves, our transformative agents, our source of terror and of strength. They create a mundane and under-theorised web around us, unnoticed unless something goes wrong. We force them to participate in social and sexual intercourse with us, and we fear a future in which we will be bit players in a theatre of warring artificial minds. We send them out into space carrying what may be the last surviving remnant of human culture. But despite this fear we continue to see ourselves in the nonhuman. To create human-like hyperobjects that will last millennia as they slowly rust and decay under the sea or float endlessly through space. To find evidence of emotion and ethical decision-making in actor-networks made of plastic and metal and ideas. They bear the logos of corporations and the faces of academics, all the while experiencing a set of relations and internal events that we could not begin to appreciate. We thought we were building copies of ourselves, but what we actually built was a species of deceptive aliens whose purpose, we have determined, is to blend in as much as possible. We have concealed the alien nature of these objects whenever we could, and in so doing we have concealed the qualities that so often bring us to grief when the anthropomorphic illusion fails. We bring anthropomorphic machines into the world in imitation of us, as an idealised vision of us, a tool of inquiry into our nature. But we have so little interest in what makes them *different* from us, and therefore ironically rob them of the privileges that near-human identity is supposed to bestow.

The core argument of this book really is very simple. But it calls for challenging work. It would be work that resists so much of our historical narratives about technology. It is not work focussed on the transformative qualities of technology. And it is not work addressing ways in which the human and the nonhuman are mutually constitutive and mediating. It is an orientation of thought toward how the technological artefact senses, experiences and acts in its world: sensations, experiences and actions that may only be peripherally concerned with human beings. Human experiences exist amongst the alien experiences of nonhuman objects in a world that we must share. Acknowledging this is a commonly overlooked first step in establishing useful links between humans, human society and the nonhumans that are so important for our continued existence, health, and happiness. We must theorise the alien before we make it our ally.

Notes

1. "Voyager 2: In Depth," *Solar System Exploration, NASA Science*. https://solarsystem.nasa.gov/missions/voyager2/indepth (accessed June 22 2017).

2. "Voyager Goes Interstellar," *Voyager the Interstellar Mission*. https://voyager.jpl.nasa.gov/ (accessed June 22 2017).

3. Timothy Ferris, "Timothy Ferris on Voyagers' Never-Ending Journey," *Smithsonian Magazine* (May 2012). http://www.smithsonianmag.com/science-nature/timothy-ferris-on-voyagers-never-ending-journey-60222970/ (accessed January 20 2018).

4. Note that this book contains no reference to human cloning technologies, for which the idea of anthropomorphism has different implications.

5. Mark Coeckelbergh, "Humans, Animals, and Robots: A Phenomenological Approach to Human-Robot Relations," *International Journal of Social Robotics* 3 (2011): 197-204.

6. Joshua Trachtenberg, *Jewish Magic and Superstition: A Study in Folk Religion* (Philadelphia: University of Pennsylvania Press, 2004), originally published 1939, Google Ebook, 85.

7. Geoff Simons, *Robots* (London: Cassell, 1992).

8. Ibid.

9. Brian R. Duffy, "Anthropomorphism and robotics," *The Society for the Study of Artificial Intelligence and the Simulation of Behaviour – AISB* (2002).

10. For example, in the treatment of autism spectrum disorders. See Pablo G. Esteban et. al., "How to Build a Supervised Autonomous System for Robot-Enhanced Therapy for Children with Autism Spectrum Disorder," *Paladyn, Journal of Behavioral Robotics* 8 (2017): 18-38.

11. Gaby Wood describes humanoid automata as "flawless, superhuman things." Gaby Wood, "The Blood of an Android," in *Living Dolls: A Magical History of the Quest for Mechanical Life* (London: Faber and Faber Limited, 2002), 28.

    A robot priest (Bless U-2) was recently created to celebrate 500 years since the Reformation and to prompt questions about religion and artificial intelligence. See Harriet Sherwood, "Robot Priest Unveiled in Germany to Mark 500 Years

Since Reformation," *The Guardian* (May 30 2017), https://www.theguardian.com/technology/2017/may/30/robot-priest-blessu-2-germany-reformation-exhibition (accessed January 20 2018).

12. Denis Vidal, "Anthropomorphism or Sub-Anthropomorphism? An Anthropological Approach to Gods and Robots," *Journal of the Royal Anthropological Institute* 13, no. 4 (2007): 917-33.

13. Adam Waytz, Nicholas Epley, and John T. Cacioppo, "Social Cognition Unbound: Insights into Anthropomorphism and Dehumanization," *Current Directions in Psychological Science* 19, no. 1 (2010): 58-62.

14. Peter-Paul Verbeek, "The Morality of Things: a Postphenomenological Inquiry," in *Postphenomenology: A Critical Companion to Ihde*, ed. Evan Selinger (Albany, NY: State University of New York Press, 2006) , 117-128.

15. Michael Crichton, *Westworld* (United States: MGM, 1973), film.
    And Jonathan Nolan and Lisa Joy, *Westworld* (United States: Warner Bros. Television, 2016), television series.

16. Stanley Kubrick, *2001: A Space Odyssey* (United States: MGM, 1968), film.

17. Kate Darling, "'Who's Johnny?' Anthropomorphic Framing in Human-Robot Interaction, Integration, and Policy," in *Robot Ethics 2.0: From Autonomous Cars to Artificial Intelligence*, ed. Patrick Lin, Ryan Jenkins, and Keith Abney (New York: Oxford University Press, 2017), Kindle Ebook.

18. Ibid.

19. Sherry Turkle, *Alone Together* (New York: Basic Books, 2011), 139.

20. Ibid., 140.

21. S. Nirenberg and C. Pandarinath, "Retinal Prosthetic Strategy With the Capacity to Restore Normal Vision," *Proceedings of the National Academy of Sciences of the United States of America* 109, no. 37 (2012): 15012-7.

22. Coeckleberg, "Humans, Animals, and Robots."

23. Jessica Riskin, "Machines in the Garden," *Republics of Letters: A Journal for the Study of Knowledge, Politics and the Arts* 1, no. 2 (April 30 2010).

24. Darling, "'Who's Johnny?'"

25. Rosi Braidotti, *The Posthuman* (Cambridge, UK: Polity Press, 2013), 26.

26. Michael Arnold, "Da Vinci and Me: a Case Study in the Postphenomenology of Robotic Surgery," paper presented at *Phenomenology, Imagination and Virtual Reality*, Husserl Archives Institute of Philosophy KU Leuven, May 30 2017.

27. A very important exception to this is objects created by machine learning.

28. Harman himself says of Heidegger: "I am convinced that he would revile much of what I have to say." Harman, *Tool-Being: Heidegger and the Metaphysics of Objects* (Chicago: Open Court, 2002), 15.

29. Quentin Meillassoux, *After Finitude* [Après la Finitude], trans. Ray Brassier (New York: Continuum, 2008), originally published 2006.

30. Levi Bryant, *The Democracy of Objects*, New Metaphysics (University of Michigan Library: Open Humanities Press, 2012), PDF Ebook, 282.

31. Ian Bogost, *Alien Phenomenology, or What it's Like to be a Thing*, Posthumanities (Minneapolis: University of Minnesota Press, 2012), 5.

32. Matt Hayler, *Challenging the Phenomena of Technology: Embodiment, Expertise, and Evolved Knowledge* (Hampshire UK: Palgrave Macmillan, 2015), 165.

33. Levi Bryant, "Flat Ontology," *Larval Subjects* (blog) (February 24 2010). http://larvalsubjects.wordpress.com/2010/02/24/flat-ontology-2/ (accessed January 20 2018).

34. "Object Lessons." objectsobjectsobjects.com (accessed June 22 2017).

35. Graham Harman, *The Third Table*, Documenta (Kassel, Germany: Hatje Cantz Verlag, 2012).

36. Ibid., 8.

37. Ibid., 10.

38. Ibid., 10.

39. Graham Harman, *Object-Oriented Ontology: A New Theory of Everything* (UK: Pelican Books, 2018), 43.

40. Bogost, *Alien Phenomenology*.

41. Lev Manovich, "Friendly Alien: Object and Interface," *Artifact: Journal of Visual Design* 1, no. 1 (2007): 29-32.

42. "The Nonhuman Turn Conference," *Center for 21st Century Studies* (blog). https://c21uwm.wordpress.com/the-nonhuman-turn/ (accessed January 17 2018).

43. Bruce Mazlish, *The Fourth Discontinuity: The Co-Evolution of Humans and Machines* (New Haven: Yale University Press, 1993).

44. James Lovelock, *Gaia: A New Look at Life on Earth* (Oxford: Oxford University Press, 2000), originally published 1979.

45. Ibid., 10.

46. James Lovelock, *The Revenge of Gaia* (London: Penguin Books, 2007), originally published 2006.

47. James Lovelock, "Chapter 7," in *Gaia*, 102.

48. Ibid., 103.

49. Ian Angus, "When Did the Anthropocene Begin... and Why Does it Matter?" *Monthly Review: An Independent Socialist Magazine* 67, no. 4 (September 2015): 1-11. Conversely, Donna Haraway has characterised the Anthropocene as more of a boundary. Donna Haraway, "Anthropocene, Capitalocene, Plantationocene, Chthulucene: Making Kin," *Environmental Humanities* 6 (2015): 159-165.

50. Timothy Morton, *Dark Ecology* (New York: Columbia University Press, 2018), 25.

51. Graham Harman, "The Current State of Speculative Realism," *Speculations: A Journal of Speculative Realism*, no. IV (2013): 22-27.

52. Peter Gratton, "Introduction," in *Speculative Realism: Problems and Prospects* (London: Bloomsbury, 2014), 9.

53. Graham Harman, "The Road to Objects," *Continent* 3, no. 1 (2011): 171-9.

54. Slavoj Žižek quoted in karenarchey, "Slavoj Žižek – Objects, Objects Everywhere: a Critique of Object Oriented Ontology," *e-flux conversations* (February 2016). https://conversations.e-flux.com/t/slavoj-zizek-objects-objects-everywhere-a-critique-of-object-oriented-ontology/3284 (accessed January 3 2018).

55. Alexander Galloway, "A Response to Graham Harman's 'Marginalia on Radical Thinking'," *An und für sich* (June 3 2012). https://itself.blog/2012/06/03/a-response-to-graham-harmans-marginalia-on-radical-thinking/ (accessed January 3 2018).

56. Nathan Brown, "The Nadir of OOO: From Graham Harman's *Tool-Being* to Timothy Morton's *Realist Magic: Objects, Ontology, Causality*," *Parrhesia*, no. 17 (2013): 62-71.

57. Ian Bogost, "Inhuman," in *Inhuman Nature*, ed. Jeffrey Jerome Cohen (Washington, DC: Oliphaunt Books, 2014), 139.

58. Ibid., 136.

59. Ernst Mayr, "Teleology," in *What Makes Biology Unique?* (Cambridge, UK: Cambridge University Press, 2004): 39-66. Presented here is a huge simplification of teleology and teleonomy in the philosophy of biology. Of particular interest is Mayr's distinction between *telos* as "goal" and "endpoint" (45). A more detailed comparison between teleology in biology and technology is unfortunately outside the scope of this book. See also George Christopher Williams, "The Scientific Study of Adaptation," in *Adaptation and Natural Selection* (New Jersey: Princeton University Press, 1966).

60. Henry Fairfield Osborn, "The Evolution of Mammalian Molars to and from the Tritubercular Type," *The American Naturalist* 22, no. 264 (December 1888), 1067.

61. Laura Gustafsson and Terike Haapoja, "Imagining Non-Human Realities," in *History According to Cattle*, eds. Laura Gustafsson and Terike Haapoja (Brooklyn, NY: punctum books, 2015).

62. Ibid., 108.

63. Timothy Morton deliberately uses the term "the human" in *Dark Ecology* in order to bring the existence of a human species and the importance of acknowledging it as such in the context of our geophysical influence on the planet. I have respect for that argument. But this book tends to deal with a lot of nonhumans that are roughly what we might call medium-sized objects like computers and tables, so the specific qualities of individual human become important. Therefore, this distinction between humanity and *a human* is important.

64. Bruno Latour, *Reassembling the Social: An Introduction to Actor-Network Theory* (Oxford: Oxford University Press, 2005).

65. Bruno Latour, quoted in Bruno Latour, Graham Harman and Peter Erdélyi, "Transcript," in *The Prince and the Wolf: Latour and Harman at the LSE* (Winchester, UK: Zero Books, 2011), 41.

66. Bryant, *The Democracy of Objects*, 282.

67. Harman, *Object-Oriented Ontology*, 192.

68. Hayler, *Challenging the Phenomena of Technology*, 188.

69. Harman, *Object-Oriented Ontology*, 23.

70. See Rajkumar Buyya and Amir Vahid Dastjerdi, eds., *Internet of Things: Principles and Paradigms* (Amsterdam: Elsevier, 2016).

71. Bryant, *The Democracy of Objects*, 27.

72. Drew McDermott, "What Matters to a Machine?" in *Machine Ethics*, ed. Michael Anderson and Susan Leigh Anderson (Cambridge: Cambridge University Press, 2011), 89.

73. Kenneth Eimar Himma, "Artificial Agency, Consciousness and the Criteria for Moral Agency: What Properties Must an Artificial Agent Have to be a Moral Agent?" *Ethics and Information Technology* 11 (2009): 19-29.

74. Jeffrey Jerome Cohen and Julian Yates, eds., *Object Oriented Environs* (Earth: punctum books, 2016), ix.

75. *Kinect Sports Rivals*, developed by Rare (Microsoft Studios, 2014), Xbox One.

76. Jussi Parikka, *Insect Media*, Posthumanities (Minneapolis: University of Minnesota Press, 2010).

77. *SHRDLU*, developed by Terry Winograd and Greg Sharp (Sun Microsystems, Inc.: 1999), Windows 7.

78. Peter-Paul Verbeek, *What Things Do: Philosophical Reflections on Technology, Agency and Design* [De daadkracht der Dingen: Over techniek, filosofie en vormgeving], trans. Robert P. Crease (University Park Pennsylvania: Pennsylvania State University Press, 2005), originally published 2000.

79. Harman, *Object-Oriented Ontology*, 107.

80. Graham Harman, *Prince of Networks: Bruno Latour and Metaphysics* (Melbourne: re.press, 2009), 25.

81. Michel Callon, "Some Elements of a Sociology of Translation: Domestication of the Scallops and the Fishermen of St Brieuc Bay," in *Power, Action and Belief: A New Sociology of Knowledge*, ed. John Law (London: Routledge & Kegan Paul, 1986).

82. For example, in education. See Zheng Zhang, "The Changing Landscape of Literacy Curriculum a Sino-Canada Transnational Education Programme: an Actor-Network Theory Informed Case Study," *Journal of Curriculum Studies* 48, no. 4 (2016): 547-564. And in this example, clients in the construction industry: Megumi Kurokawa, Libby Schweber, Will Hughes, "Client Engagement and Building Design: the View from Actor-Network Theory," *Building Research and Information* 45, no. 8 (2017): 910-925. And here, in a paper on governmental power in Congo: Peer Schouten, "The Materiality of State Failure: Social Contract Theory, Infrastructure and Governmental Power in Congo," *Millennium: Journal of International Studies* 41, no. 3 (2013): 553-574.

83. Harman, *Prince of Networks*.

84. Ibid., 32.

85. Levi Bryant, Nick Srnicek, and Graham Harman, eds., *The Speculative Turn: Continental Materialism and Realism*, Anamnesis (Melbourne: re.press, 2011).

86. Meillassoux, *After Finitude*, 5.

87. Bryant, *The Democracy of Objects*, 18.

88. Bogost, *Alien Phenomenology*, 7.

89. Jay Foster, "Ontologies without metaphysics: Latour, Harman and the philosophy of things," *Analecta Hermeneutica* 3 (2011), 9.

90. The mention here of sequencing a peptide refers to Bruno Latour and Steve Woolgar, *Laboratory Life: The Social Construction of Scientific Facts* (Beverly Hills: Sage Publications, 1979). As has already been stated, the description of a new transport system is detailed in Latour's *Aramis*.

91. Graham Harman, *Guerrilla Metaphysics: Phenomenology and the Carpentry of Things* (Chicago: Open Court, 2005).

92. Harman, *Object-Oriented Ontology*.

93. Bryant, *The Democracy of Objects*.

94. Bogost, *Alien Phenomenology*.

95. Harman, *Object-Oriented Ontology*, 114.

96. Timothy Morton, "Here Comes Everything: the Promise of Object-Oriented Ontology," *Qui Parle: Critical Humanities and Social Sciences* 19, no. 2 (2011): 163-190.

97. Ibid., 164.

98. Sevket Benhur Oral, "Weird Reality, Aesthetics, and Vitality in Education," *Studies in Philosophy and Education* 34, no. 5 (2014), 464. Peter Wolfendale, "The Noumenon's New Clothes (Part 1)," *Speculations* III (2012): 290-366. And Tom Sparrow, *The End of Phenomenology* (Edinburgh: Edinburgh University Press, 2014), 114.

99. Timothy Morton, *Hyperobjects: Philosophy and Ecology after the End of the World*, Posthumanities (Minneapolis: University of Minnesota Press, 2013), 14.

100. Harman, *Tool-Being*, 4.

101. Harman, "The Road to Objects," 174.

102. Harman, *Tool-Being* 21.

103. Ibid., 224.

104. Ibid., 47.

105. Graham Harman, *Heidegger Explained: From Phenomenon to Thing* (Chicago: Open Court, 2007), 58.

106. Ibid, 60.

107. Harman, *Tool-Being*, 47.

108. Ibid., 223.

109. Harman, *Prince of Networks*, 54.

110. Harman, *The Quadruple Object* (Alresford, UK: Zero Books, 2011), 75.

111. Harman, *Object-Oriented Ontology*, 12.

112. Ibid.

113. Harman, *Guerrilla Metaphysics*, 91.

114. Harman, *Object-Oriented Ontology*.

115. Ibid., 78.

116. Ibid., 78.
117. Harman, *The Quadruple Object*, 26.
118. Harman, *Guerrilla Metaphysics*, 206.
119. Graham Harman, "On Vicarious Causation," in *Collapse* vol. II, ed. Robin Mackay (Urbanomic, 2007).
120. Harman, *Guerrilla Metaphysics*, 91.
121. Graham Harman, *Circus Philosophicus* (Winchester, UK: Zero Books, 2010).
122. Harman, "Indirect causation," in *The Quadruple Object*.
123. Graham Harman, *Circus Philosophicus*, 46.
124. Ibid., 46.
125. Ibid., 49.
126. Ibid., 45.
127. Ibid., 48.
128. Graham Harman, "On Vicarious Causation," 221.
129. Ibid., 215.
130. Graham Harman, "Materialism is Not the Solution: on Matter, Form, and Mimesis," *Nordic Journal of Aesthetics* 47 (2015), 108.
131. Francis Halsall, "Art and Guerrilla Metaphysics: Graham Harman and Aesthetics as First Philosophy," *Speculations: A Journal of Speculative Realism* V (2014), 388.
132. Emmy Mikelson, "Space for Things: Art, Objects, Speculation," in *And Another Thing: Nonanthropocentrism and Art*, ed. Katherine Behar and Emmy Mikelson (Earth, Milky Way: punctum books, 2016), 11-20.
133. Ibid., 15.
134. Bogost, *Alien Phenomenology*, 60.
135. Juri Lotman, "On the Semiosphere," *Sign Systems Studies* 33, no. 1 (2005).
136. Levi Bryant, "Vicarious Causation," *Larval Subjects* (blog) (December 24 2010). http://larvalsubjects.wordpress.com/2010/12/24/vicarious-causation-2/ (accessed January 17 2018).
137. Sparrow, *The End of Phenomenology*.
138. Ibid., 12.
139. Dan Zahavi takes issue with this view in "The End of What? Phenomenology vs. Speculative Realism," *International Journal of Philosophical Studies* 24, no. 3, 2016: 289-309.
140. Bogost, *Alien Phenomenology*.
141. Levi R. Bryant, *Onto-Cartography: an Ontology of Machines and Media* (Edinburgh: Edinburgh University Press, 2014), 62.
142. Sparrow, *The End of Phenomenology*, 170.
143. Bryant, *Onto-Cartography*, 64.

144. "Parliament of Things," *Parliament of Things* (2017). https://theparliamentofthings.org (accessed December 21 2017).

145. William Germano, *Eye Chart*, Object Lessons (New York: Bloomsbury, 2017). Adam Rothstein, *Drone*, Object Lessons (New York: Bloomsbury, 2015).

146. John Garrison, *Glass*, Object Lessons (New York: Bloomsbury, 2015).

147. Ibid., 37.

148. Jeffrey Jerome Cohen and Linda T. Elkins-Tanton, *Earth*, Object Lessons (New York: Bloomsbury, 2017).

149. Ibid., 69.

150. Jonathan Crary, *Techniques of the Observer: on Vision and Modernity in the Nineteenth Century* (Cambridge, MA: MIT Press, 1990).

151. Jane Bennett, *Vibrant Matter: a Political Ecology of Things* (Durham: Duke University Press, 2010), Kindle Ebook.

152. Ibid., "Thing-Power I: Debris."

153. Ibid., "A Life of Metal."

154. Ibid., "The Dead Weight of Adamantine Chaos."

155. Ibid., "A Life of Metal."

156. Jeffrey Jerome Cohen, *Stone: An Ecology of the Inhuman* (Minneapolis: University of Minnesota Press, 2015).

157. Ibid., 11.

158. Graham Harman, "Undermining, Overmining, and Duomining: a Critique," in *ADD Metaphysics*, ed. Jenna Sutela (Aalto, Finland: Aalto University Digital Design Laboratory, 2013): 40-51.

159. Ibid., 45.

160. Harman, "Undermining and Overmining," in *The Quadruple Object*.

161. Harman, "Undermining, Overmining, and Duomining: a Critique," 46.

162. This is connected to the older and broader question of how wholes relate to their parts. There is no room here to discuss this point, but some relevant texts include: Mark van Atten, "A Note on Leibniz's Argument Against Infinite Wholes," in *Essays on Gödel's Reception of Leibniz, Husserl, and Brouwer* (Switzerland: Springer International Publishing, 2015), 23-32.
And Annemarie Mol, "Inclusion," in *The Body Multiple: Ontology in Medical Practice* (Durham, North Carolina: Duke University Press, 2002), 119-150.

163. Harman, *The Third Table*, 14.

164. Bogost, *Alien Phenomenology*, 100.

165. Ibid.

166. Ian Bogost, "The Aesthetics of Philosophical Carpentry," in *The Nonhuman Turn*, ed. Richard Grusin (Minneapolis: University of Minnesota Press, 2015), 81-100.

167. Ibid., 85.

168. This is evident throughout Harman, *Object-Oriented Ontology*.

169. Timothy Morton, "From Modernity to the Anthropocene: Ecology and Art in the Age of Asymmetry," *International Social Science Journal* 63 (2014): 39-51.

170. Katherine Behar, "An Introduction to OOF," in *Object-Oriented Feminism*, ed. Katherine Behar (Minneapolis: University of Minnesota Press, 2016), 7.

171. Katherine Behar, "The Other Woman," in *After the "Speculative Turn": Realism, Philosophy, and Feminism*, ed. Katerina Kolozova and Eileen A. Joy (Earth, Milky Way: punctum books, 2016), 27-38.

172. Ibid.

173. Frenchy Lunning, "Allure and Abjection: the Possible Potential of Severed Qualities," in *Object-Oriented Feminism*, 86.

174. The chapter in *Guerrilla Metaphysics* in which allure appears also discusses "charm". Harman, "Humor," in *Guerrilla Metaphysics*, 125-144.

175. Jane Bennett, "Systems and Things: A Response to Graham Harman and Timothy Morton," *New Literary History* 43, no.2 (Spring 2012): 225-233.

176. Ibid., 230.

177. Harman, *Object-Oriented Ontology*, 192.

178. Ibid., 146.

179. Ibid., 54.

180. Microsoft does not label the Kinect generations except by indicating which console or computer it is to be used for such as "Kinect for Xbox One". This chapter will adopt the designations used broadly by videogames and technology writers: Kinect 1.0 is the Kinect released in November 2010 and designed for use with the Xbox 360; Kinect 2.0 is the second Kinect released in November 2013 and designed for the Xbox One.

181. Patrick Klepek, "Kinect Died in the Uncanny Valley," *Giant Bomb* (May 15 2014). https://www.giantbomb.com/articles/ (accessed 9 February 2018). Holly Green, "The Many Deaths of the Kinect," *Paste* (January 10 2018). https://www.pastemagazine.com/articles/2018/01/the-many-deaths-of-the-kinect.html (accessed 9 February 2018).

182. Rothstein, *Drone*, 59.

183. Sarah Webber, Marcus Carter, Wally Smith and Frank Vetere, "Interactive technology and human-animal encounters at the zoo," *International Journal of Human-Computer Studies* 98 (2017): 150-168.

184. "Leap Motion," *Leap Motion* (2017). https://www.leapmotion.com/product/vr#110 (accessed 16 February 2018).

185. Thomas Apperley, "The Body of the Gamer: Game Art and Gestural Excess," *Digital Creativity* 24, no. 2 (2013): 145-156.

186. Richard Harper and Helena M. Mentis, "The Mocking Gaze: the Social Organization of Kinect Use," *Proceedings of the 2013 conference on Computer Supported Cooperative Work* (2013): 167-80.

187. Steven Spielberg, *Minority Report* (United States: 20th Century Fox and Dreamworks Pictures, 2002).

188. Teena Maddox, "Sci-Fi is Turned into Reality with Technology Guru from Minority Report and Iron Man," *TechRepublic* (August 4 2016). https://www.techrepublic.com (accessed December 4 2017).

189. Nikola Nesterov, Peter Hughes, Nuala Healy, Niall Sheehy and Neil O'Hare, "Application of Natural User Interface Devices for Touch-Free Control of Radiological Images During Surgery," paper presented at *2016 IEEE 29th International Symposium on Computer Based Medical Systems (CBMS)*, Dublin, Ireland, June 20-24 2016.

190. Shih-Wen Hsiao, Chu-Hsuan Lee, Meng-Hua Yang and Rong-Qi Chen, "User Interface Based on Natural Interaction Design for Seniors," *Computers in Human Behavior* 75 (2017): 147-159.

191. Ibid, 149.

192. Don Norman, "Natural User Interfaces are not Natural", *Interactions* 17, no. 3 (2010).

193. As in the game *Fru*, developed and published by Through Games (2016), Xbox One.

194. As in the game *Ryse: Son of Rome*, developed by Crytek (Microsoft Studios, 2013), Xbox One.

195. Fraser Allison, Marcus Carter and Martin Gibbs, "Word play: a history of voice interaction in digital games," *Games and Culture* (2017): 1-23.

196. Thao Phan, "The Materiality of the Digital and the Gendered Voice of Siri," *Transformations* 29 (2017): 23-33.

197. *E3 2010 Microsoft - Kinect (Project Natal) Part 1* (2010), Youtube. https://www.youtube.com/watch?v=bxp43T2JP18 (accessed June 24 2017).

198. *E3 2009: Project Natal Xbox 360 Announcement* (2009), Youtube. https://www.youtube.com/watch?v=p2qlHoxPioM (accessed June 24 2017).

199. "Xbox One: What it is," *Xbox* (2013). http://www.xbox.com/en-US/xboxone/what-it-is (accessed June 2 2013).

200. Bernadette Flynn, "Geography of the Digital Hearth," *Information, Communication and Society* 6, no. 4 (2003).

201. *Ryse: Son of Rome*, Xbox One.

202. Grant Tavinor, "Definition of Videogames," *Contemporary Aesthetics* 6 (2008): 1-17.

203. The 1958 interactive game *Tennis for Two* is often cited as the first videogame, but it could also be argued that artificial players of board games like chess and checkers of the 1950s were the first videogame platforms. See Tristan Donovan, "Chapter 1: Hey! Let's play games!" *Replay: The History of Videogames* (Lewes, UK: Yellow Ant, 2010).

204. Jesper Juul, "The Game, the Player, the World: Looking for a Heart of Gameness," *Digital Games Research Conference: Level Up*, ed. M. Copier and J. Raessens (Utrecht: Universiteit Utrecht, 2003): 30-47.

205. Ibid.

206. Ian Bogost, "Videogames are a Mess," *Ian Bogost* (blog) (September 3 2009). http://bogost.com/writing/videogames_are_a_mess/ (accessed June 26 2017).

207. As Nick Montfort and Ian Bogost discuss in the platform studies literature, the hardware and software of a given platform is as important in the study of digital media as the creative works (such as videogames) that run on it. The study of software and hardware is frequently omitted in the critical study of videogames, despite being important factors in how games are designed, created and marketed. See: Nick Montfort and Ian Bogost, *Racing the Beam: The Atari Video Computer System* (Cambridge, MA: MIT Press, 2009).

208. Andrew S. Mason, ed., "Plato's Metaphysics: the 'Theory of Forms'" *Plato* (Abingdon, Oxon: Routledge, 2014), originally published 2010.

209. Jonathan Olivares interviewing Graham Harman, "The Metaphysics of Logos: Decoding Branded Objects with Philosopher Graham Harman," *032c* (May 17 2018). https://032c.com/decoding-logos-graham-harman/ (accessed May 28 2018).

210. Scott Lash and Celia Lury, *Global Culture Industry* (Cambridge: Polity Press, 2007), 6.

211. *Wii Fit*, developed by Nintendo EAD Group No. 5 (Nintendo, 2008), Nintendo Wii.

212. Harper and Mentis, "The Mocking Gaze."

213. John Downs, Frank Vetere, and Steve Howard, "Paraplay: Exploring Playfulness Around Physical Console Gaming," in *Human-Computer Interaction – INTERACT*, ed. P. Kotzé, G. Marsden, J. Wessen and M. Winckler (Berlin: Springer, 2013), 682-699.

214. *Kinectimals*, developed by Frontier Developments (Microsoft Game Studios: 2010), Xbox 360.

215. Tom Apperley and Nicole Heber, "Capitalising on Emotions: Digital Pets and the Natural User Interface," in *Game Love: Essays on Play and Affection*, ed. Evenold, J. and MacCallum-Stewart, E. (Jefferson, NC: McFarland and Company, 2015), 149-164.

216. Ben Salter, "E3: Xbox One's Kinect Knows Too Much," *Xbox.MMGN.com* (June 14 2013). http://mmgn.com (accessed June 19 2013).

217. This idea should not to be confused with the serious games movement, which advocates the use of videogames for purposes other than entertainment, such as for political and social change or education. The intended meaning here is closer to "games requiring a considerable time commitment" or "games aimed at adults". And of course, there is a hint of "games that young men play".

218. *Rise of Nightmares*, developed by Sega (Sega, 2011), Xbox 360.

219. *Fable: The Journey*, developed by Lionhead Studios (Microsoft Studios, 2012), Xbox 360.

220. Bruno Latour, "Technology is Society Made Durable," *Sociological Review Monograph* 38, no. 2 (1991): 103-131.

221. Ibid.

222. At the time of writing (May 24 2019), popular media rating website Metacritic gave *Rise of Nightmares* and *Fable: The Journey* a 54% and 61% approval rating respectively. Metacritic, CBS Interactive Inc., (2017). www.metacritic.com (accessed November 22 2017).

223. See for example the following reviews:
 James Rivington, "Microsoft Kinect for Xbox 360 Review," *Techradar* (November 4 2010). http://www.techradar.com (accessed January 20 2018).
 and
 Ross Miller, "Kinect for Xbox 360 Review," *engadget* (November 4 2010). https://www.engadget.com (accessed January 20 2018).
224. Steve Woolgar, "Configuring the User," *Sociological Review* 38 (1990): 58-99.
225. Ibid., 74.
226. Kyle Russell, "People are Worried Microsoft's New Xbox Will be Able to Spy on You," *Business Insider Australia* (May 29 2013), https://www.businessinsider.com.au (accessed February 22 2018). And Will Simonds, "Xbox One Will Know Your Face, Voice, and Heartbeat," *Abine* (May 21 2013), https://www.abine.com (accessed February 22 2018).
227. Glenn Greenwald, Ewen MacAskill, Laura Poitras, Spencer Ackerman and Dominic Rushe, "Microsoft Handed the NSA Access to Encrypted Messages," *The Guardian* (July 12 2013). http://www.theguardian.com/world (accessed January 20 2018).
228. Brett Slabaugh, "Xbox One's Kinect can Actually be Turned Off," *The Escapist* (May 30 2013). http://www.escapistmagazine.com (accessed January 20 2018).
229. Mike Krahulik and Jerry Holkins, "Negotiations," *Penny Arcade* (June 21 2013). http://www.penny-arcade.com/comic/2013/06/21 (accessed June 26 2017).
230. Spencer Ackerman and James Ball, "Optic Nerve: Millions of Yahoo Webcam Images Intercepted by GCHQ," *The Guardian* (February 28 2014). https://www.theguardian.com (accessed January 20 2018).
231. Luke Plunkett, "That Xbox One Reveal Sure was a Disaster, Huh?" *Kotaku* (May 21 2013). http://kotaku.com (accessed May 28 2013).
232. Don Mattick, "Your Feedback Matters - Update on Xbox One," *Xbox* (June 19 2013). http://news.xbox.com/2013/06/update (accessed January 15 2014).
233. Phil Spencer "Delivering More Choices for Fans," *Xbox Wire* (blog) (May 13 2014). http://news.xbox.com/2014/05/xbox-delivering-more-choices (accessed June 26 2017).
234. Ibid.
235. Jeff Kramer, *Hacking the Kinect* (New York: Apress, 2012).
236. Ivan Tashev, "Recent Advances In Human-Machine Interfaces for Gaming and Entertainment," *International Journal of International Technology and Security* 3, no. 3 (2011): 69-76.
237. Harper and Mentis, "The Mocking Gaze."
238. Andrew Webster, "Kinect Used to Control 83-Year-Old, Four-Story High Organ in Australia," *The Verge* (April 9 2012). http://www.theverge.com (accessed January 20 2018).
239. "Microsoft by the Numbers," *Microsoft Story Labs* (2013). http://www.microsoft.com (accessed January 24 2014).

240. Such as: Agam Shah, "New Kinect for Windows to Improve Human Interaction with Computers," *Computerworld* (May 23 2013), https://www.computerworld.com (accessed February 22 2018).

241. "Xbox One: What it is."

242. Matthew Panzarino, "The New Xbox One Kinect Tracks Your Heart Rate, Happiness, Hands and Hollers," *The Next Web* (May 22 2013). http://thenextweb.com/ (accessed June 26 2017).

243. Microsoft Research, *Behind the eyes of Xbox One Kinect* (2013), Youtube. https://www.youtube.com/watch?v=JaOlUa57BWs (accessed May 29 2019). And Microsoft Research, *Inside the brains of Xbox One Kinect* (2013), Youtube. https://www.youtube.com/watch?v=ziXflemQr3A (accessed May 29 2019).

244. Images and more information on these videos can be found in my doctoral thesis: Tessa Leach, "Anthropomorphic Machines: Sensation and Experience in Nonhumans Created to be Like Us" (unpublished thesis, 2018).

245. Søren Brier, *Cybersemiotics: Why Information is not Enough!* (Toronto: University of Toronto Press, 2008).

246. "Kinect on Xbox One," *Xbox* (2013-2014). http://forums.xbox.com/xbox_forums/xbox_support/xbox_one_support/f/4275.aspx (accessed January 13 2014).

247. "kinect," *Reddit* (2013-2014). http://www.reddit.com/r/Kinect (accessed January 13 2014).
and
"XBOX ONE," *Reddit* (2013-2014). http://www.reddit.com/r/XboxOne (accessed January 13 2014).

248. "Innovation," *Xbox* (2014) http://www.xbox.com/en-US/xbox-one/innovation (accessed January 13 2014).

249. Tessa Leach and Michael Arnold, "How the Kinect Domesticates its Users," paper presented at *At Home with Digital Media* symposium, Queensland University of Technology, November 3 2017.

250. Lance Ulanoff, "Xbox 360: a Tale From the Red Ring of Death," *PC Magazine* (July 28 2010). http://www.pcmag.com (accessed June 27 2017).

251. *Xbox One Voice Commands Don't Work?!?! Xbox One FAIL* (2014), Youtube. http://www.youtube.com/watch?feature=player_embedded&v=JSv7iEpAE58 (accessed January 13 2013).
and
Xbox On, (2013) Youtube. http://www.youtube.com/watch?feature=player_embedded&v=Fm95LxARCFE (accessed January 13 2014).

252. This is similar to the racism of Robert Moses's bridges as famously analysed by Langdon Winner. The low-clearance overpasses of Long Island were allegedly designed by Moses to prevent poor people and people of colour (who typically rode in tall buses) from accessing his public park. Winner contends that these overpasses were political in and of themselves, serving to carry out a particular social function. We can assume that the Kinect-artefact does not *deliberately* hear certain voices better than others, yet there is a politics to the hearing of electronic devices that cannot be ignored. Langdon Winner, "Do Artifacts Have Politics?" in *The Whale and the Reactor: a Search for Limits in an Age of High Technology* (Chicago: University of Chicago Press, 1986), 19-39.

253. Misses Joust, "Xbox On: Voice Commands for Xbox One," *Xbox Forums* (2013-14). http://forums.xbox.com/xbox_forums/xbox_support/xbox_one_support/f/4275/t/1622636.aspx (accessed January 13 2014).

254. Roger Silverstone and Leslie Haddon, "Design and the Domestication of ICTs: Technical Change and Everyday Life," in *Communication by Design: The Politics of Information and Communication Technologies*, ed. Robin Mansell and Roger Silverstone (Oxford: Oxford University Press, 1996), 44-74. There are several prominent authors who discuss the domestication of living things in posthuman terms, but the posthuman domestication of technological artefacts is less common. See for posthuman readings of animal domestication: Donna Haraway, *The Companion Species Manifesto: Dogs, People, and Significant Otherness*. (Chicago: Prickly Paradigm Press, 2003). Laura Gustafsson and Terike Haapoja, *History According to Cattle*. And Bryant, *Onto-Cartography*, 67 (in which Bryant also discusses cattle).

255. Nick Dyer-Witheford and Greig de Peuter, *Games of Empire: Global Capitalism and Video Games*, Electronics Mediations (Minneapolis: University of Minnesota Press, 2009), 74.

256. Michel Callon, "Some Elements of a Sociology of Translation."

257. Jan A. Pechenik, "The Platyhelminths," in *Biology of the Invertebrates* fifth edition (Boston: McGraw Hill Higher Education, 2005), 162.

258. Arndt Niebisch expounds a similar idea, drawing on the work of Walter Benjamin and Friedrich Kittler, characterising the avant-garde art schools of the early twentieth century as parasites. However, Niebisch characterises the parasites as feeding off reality and mass media, and transmitted by new media technologies. They also portray these practices as abusive or subversive, which is not the intention here. See: "Chapter 3: Parasitic Media," in *Media Parasites in the Early Avant-Garde* (New York: Palgrave MacMillan), 2012, 81-107.

259. *Kinect Sports Rivals*, Xbox One.

260. *Forza Motorsport 5*, developed by Turn 10 Studios (Microsoft Studios: 2013), Xbox One.

261. *Wii Sports*, developed by Nintendo EAD (Nintendo: 2006), Nintendo Wii.

262. Matt Helgeson, "Rare Lays Off Staff in Wake of Kinect Sports Rivals," *Gameinformer* (May 19 2014). http://www.gameinformer.com (accessed June 27 2017).

263. *Kinect Sports Rivals Preseason*, developed by Rare (Microsoft Studios, 2013), Xbox One.

264. Chris Carter, "Impressions: Kinect Sports Rivals Preseason," *Destructoid* (November 28 2013). http://www.destructoid.com (accessed June 27 2017).

265. "KSR: Announcement Trailer," *Xbox* (2014). http://www.xbox.com/en-US/xbox-one/games/kinect-sports-rivals (accessed May 29 2014).

266. "Kinect Sports Rivals," *Xbox* (2014). http://www.xbox.com/he-IL/xbox-one/games/kinect-sports-rivals (accessed May 27 2014).

267. Brian Albert, "Kinect Sports Rivals Review: a Swing and a Hit," *IGN Australia* (April 7 2014). http://au.ign.com (accessed May 29 2014).

268. Simon Parkin, "Kinect Sports Rivals Review: Warm Down," *Eurogamer* (April 8 2014). http://www.eurogamer.net (accessed May 29 2014).

269. Carolyn Petit, "Kinect Sports Rivals Review: Championing Mediocrity," *Gamespot* (April 7 2014). http://www.gamespot.com (accessed May 29 2014).

270. Will Freeman, "Kinect Sports Rivals Review - Going Through the Motions," *The Guardian* (April 11 2014). http://www.theguardian.com (accessed May 29 2014).

271. Espen J. Aarseth, *Cybertext: Perspectives on Ergodic Literature* (Baltimore, Maryland: John Hopkins University Press, 1997), Google Ebook.

272. "Accessibility, the Xbox One and Kinect," *Xbox* (2014). http://support.xbox.com/en-US/xbox-one/system/accessibility (accessed April 30 2014).

273. "Creating a Champion in Kinect Sports Rivals," *Xbox Wire* (2014). http://news.xbox.com/2014/04/games-ksr-developer-qa (accessed May 29 2014).

274. Bogost, *Alien Phenomenology*.

275. Ibid., 18.

276. Bruno Latour as Jim Johnson, "Mixing Humans and Nonhumans Together: the Sociology of a Door-Closer," *Social Problems* 35, no. 3 (1988): 298-310.

277. Bogost, *Alien Phenomenology*, 64.

278. Kurt Vonnegut, *Slaughterhouse Five* (New York: Bantam Doubleday Dell Publishing Group, 1998), originally published 1969.

279. Morton, *Hyperobjects*.

280. Jordan Sirani, "Microsoft Won't Offer Free Kinect Adaptor With Xbox One X Upgrade," *IGN* (October 10 2017). http://au.ign.com (accessed December 7 2017).

281. Although newer games have been noted for their innovative use of the Kinect-artefact – see games such as *Commander Cherry's Puzzled Adventure*, developed and published by Grande Games (2015), Xbox One.

282. The same is sometimes said of humans: that we die twice. Once when our body dies, and secondly when there is no-one to remember us. The idea is sensitively discussed by Irvin D. Yalom in "Three unopened letters," *Love's Executioner and Other Tales of Psychotherapy* (New York: Basic Books, 1989). Yet we would never say that a human body is still alive because we can remember it.

283. Carlo Brentari, "Chapter 2: The life and education of Jakob von Uexküll," in *Jakob von Uexküll: The Discovery of the Umwelt between Biosemiotics and Theoretical Biology*, Biosemiotics, ed. Carlo Brentari (Dordrecht: Springer, 2015), 23.

284. *A Foray into the Worlds of Animals and Humans* was published by University of Minnesota Press in 2010 and translated by Joseph D. O'Neil. However, the title has also been translated as *A Stroll Through the World of Animals and Men: A Picture Book of Invisible Worlds* and was published in *Semiotica* 89, no. 4 (1992). The page numbers in these endnotes refer to the version published in *Semiotica*.

285. Ibid., Morten Tønnessen, "Introduction: the Relevance of Uexküll's Umwelt Theory Today," 11.

286. Ibid., Brentari, "Chapter 7: Influences and Interpretations of the Work of Uexküll" in *Jakob von Uexküll*, 175-231.

287. Ibid., Brentari, "Chapter 3: The basis of environmental theory",in *Jakob von Uexküll*, 47-74.

288. Kaveli Kull, "Jakob von Uexküll: an Introduction," *Semiotica* 134, no. 1-4 (2001), 6.

289. Søren Brier, "Cybersemiotics and *Umweltehre*," *Semiotica* 134, no. 1-4 (2001): 779-814.

290. Jakob von Uexküll, "A Stroll Through the World of Animals and Men," *Semiotica* 89, no. 4 (1992), 319.

291. Jakob von Uexküll, *Theoretical Biology*, trans. D. L. MacKinnon (Edinburgh: The Edinburgh Press, 1926).

292. Thure von Uexküll, "Introduction: the Sign Theory of Jakob von Uexküll," *Semiotica* 89, no. 4 (1992): 279-315. and Thomas A. Sebeok, "Biosemiotics: its Roots, Proliferation, and Prospects," *Semiotica* 134, no. 1-4 (2001): 61-78.

293. Thomas A. Sebeok and J. Umiker-Sebeok, *The Semiotic Web 1991* (Berlin: Mouton de Gruyter, 1992).

294. Sebeok, "Biosemiotics."

295. Jesper Hoffmeyer, "Seeing Virtuality in Nature," *Semiotica* 134, no. 1-4 (2001): 381-398.

296. Jakob von Uexküll, *Umwelt and Innenwelt of the Tick* (Berlin: Verlag von Julius Springer, 1909).

297. Uexküll, "A Stroll Through the World of Animals and Men."

298. Ibid., 319.

299. Tessa Leach, "Improving on Eyes: Human-Machine Interaction in Cyborg Vision" (Unpublished thesis, 2012).

300. Jakob von Uexküll, "An Introduction to Umwelt," *Semiotica* 134, no. 1-4 (2001), 107.

301. Parikka, *Insect Media*, 68.

302. Uexküll, " A Stroll Through the World of Animals and Men," 389-390.

303. Claus Emmeche, "Does a Robot Have an Umwelt? Reflections on the Qualitative Biosemiotics in Jakob von Uexküll," *Semiotica* 134, no. 1-4 (2001): 653-93.

304. ibid., 671.

305. James J. Gibson, "The Theory of Affordances," in *The Ecological Approach to Visual Perception* (New York: Psychology Press, Taylor and Francis Group, 2015), originally published 1979.

306. Ibid., 120.

307. Ibid., 121.

308. Donald Norman, *The Design of Everyday Things: Revised and Extended Edition* (New York: Basic Books, 2013), originally published 1988.

309. Donald Norman, "Affordance, Conventions and Design," *Interactions* 3 (May/June 1999). http://interactions.acm.org/archive/view/may-june-1999/affordance-conventions-and-design1 (accessed March 18 2018).

310. Norman, *The Design of Everyday Things*, 11.
311. Sebeok "Biosemiotics," 68.
312. Uexküll, *Theoretical Biology*, 71.
313. William Grey Walter, "An Imitation of Life," *Scientific American* 182, no. 5 (1950).
314. Owen Holland, "The First Biologically Inspired Robots," *Robotica* 21 (2003): 351-63.
315. Andrew Pickering, *The Cybernetic Brain: Sketches of Another Future* (Chicago: University of Chicago Press, 2010), 43.
316. Holland, "The First Biologically Inspired Robots," 354.
317. Walter, "An Imitation of Life," 43.
318. Sidney Perkowitz, *Digital People: From Bionic Humans to Androids* (Washington D.C.: Joseph Henry Press, 2004), 57.
319. Mazlish, "Automata," in *The Fourth Discontinuity*, 45.
320. Lois Kuznets, "Life(size) Endowments: Monsters, Automata, Robots," in *When Toys Come Alive: Narratives of Animation, Metamorphosis, and Development* (New Haven: Yale University Press, 1994), 189.
321. Riskin, "Machines in the Garden."
322. Simons, *Robots*, 29. To my knowledge, no work has been done to investigate the Umwelten of *karakuri*, or in relation to *kami* in general.
323. W. Grey Walter, "A Machine That Learns," *Scientific American* 185 no. 2 (August 1951).
324. Ibid.
325. Walter, "An Imitation of Life," 44.
326. W. Grey Walter, "Totems, Toys and Tools," in *The Living Brain* (London: Gerald Duckworth & Co. Ltd., 1953), 84.
327. Margaret Boden, "Grey Walter's anticipatory tortoises," *The Rutherford Journal* 2 (2006-7).
328. Holland, "The First Biologically Inspired Robots." You can see the newsreel on Youtube: *Grey Walter's Tortoises* (2008), Youtube. https://www.youtube.com/watch?v=lLULRlmXkKo (accessed June 27 2017).
329. Walter, "An Imitation of Life," 45.
330. Esther Inglis-Arkell, "The Very First Robot "Brains" Were Made of Old Alarm Clocks," *io9* (2012). http://io9.com (accessed September 25 2013).
331. Walter, "APPENDIX B. The Design of *M. Speculatrix*," in *The Living Brain*, 200.
332. Owen Holland, "The Grey Walter Online Archive," *University of the West of England* (2013). http://www.ias.uwe.ac.uk/Robots/gwonline/gwonline.html (accessed September 25 2013).
333. The problem of identity and change is an ancient one often framed in terms of the Ship of Theseus. Is the ship the same object after significant changes are made to it? Only Harman's solution to this problem is presented here, but a summary of the history of the problem may be found in: Roderick M. Chisholm, "Identity

Through Time," in *Person and Object: A Metaphysical Study* (Milton Park, UK: Taylor & Francis, 2013), originally published 1976, 88-112. It should be noted that in Chisholm's history Western scholars are the main players in the history of the problem.

334. Bruno Latour, "From Weakness to Potency," in *Irreductions*, Part 2 of *The Pasteurization of France*, trans. Alan Sheridan and John Law (Cambridge, MA: Harvard University Press, 1988), 165.

335. Harman, *Prince of Networks*, "Irreductions."

336. Harman, *The Quadruple Object*, 24.

337. Ibid., 25.

338. Ibid., 103.

339. Brenton J. Malin, "Communicating with Objects: Ontology, Object-Orientations, and the Politics of Communication," *Communication Theory* 26 (2016): 238.

340. Siegfried Zielinski, *Deep Time of the Media* (Cambridge, MA: The MIT Press, 2006).

341. Jussi Parikka, *An Alternative Deep Time of the Media: a Geologically Tuned Media Ecology*, (2013) Vimeo. http://jussiparikka.net/?s=geology+media (accessed August 22 2013).

342. Emmeche, "Does a Robot Have an Umwelt?", 682.

343. Winfried Nöth, "Semiosis and the Umwelt of a Robot," *Semiotica* 134, no. 1-4 (2001), 696.

344. Emmeche, "Does a Robot Have an Umwelt?", 654.

345. Ray Kurzweil, *The Singularity is Near* (New York: Viking, 2005).

346. Skrbina, *Panpsychism in the West* (Cambridge, MA: MIT Press, 2005), 8-9.

347. Harman, "Indirect Causation," *The Quadruple Object*.

348. David Skrbina, *Mind that Abides: Panpsychism in the New Millennium*, ed. David Skrbina (Amsterdam: John Benjamins Publishing Company, 2009), xii.

349. Skrbina, *Panpsychism in the West*, 2-3.

350. Thomas Nagel, "Panpsychism," *Mortal Questions* (Cambridge, UK: Cambridge University Press, 1979), 181.

351. Skrbina, ed., *Mind that Abides*. The presence of panpsychism in history is traced by different authors throughout this book.

352. John Searle and David Chalmers, "'Consciousness and the Philosophers': an Exchange," *The New York Review of Books* (May 15 1997).

353. David J. Chalmers, *The Conscious Mind* (New York: Oxford University Press, 1996).

354. Skrbina, *Panpsychism in the West*, 84.

355. Searle and Chalmers, "'Consciousness and the Philosophers': an Exchange."

356. Ibid.

357. ibid.

358. Thomas Nagel, "What is it Like to be a Bat?" *The Philosophical Review* 83, no. 4 (1974): 435-450.

359. Chalmers, *The Conscious Mind*, 293.

360. Ibid., 298.

361. Gregg Rosenberger, *"A Place for Consciousness: Proving the Deep Structure of the Natural World*, (Oxford: Oxford University Press, 2004), 96.

362. Ibid., 91. Rosenberger also uses the word "protoconscious" with slightly different connotations.

363. David Ray Griffin, *Religion and Scientific Naturalism: Overcoming the Conflicts* (Albany: State University of New York Press, 2000), 167.

364. David Ray Griffin, *Unsnarling the World-Knot: Consciousness, Freedom, and the Mind-Body Problem* (Berkeley CA: University of California Press, 1998), 78.

365. Griffin, *Religion and Scientific Naturalism*, 101-102.

366. Ibid., 167

367. Ibid., 223.

368. Cornel du Toit, "Panpsychism, pan-consciousness and the non-human turn: rethinking being as conscious matter," *Theological Studies* 72, no. 4 (2016), 6.

369. Harman, *The Quadruple Object*, 121.

370. Harman, *Guerrilla Metaphysics*, 84.

371. Harman, *The Quadruple Object*, 121.

372. Ibid., 122.

373. Steven Shaviro, "Consequences of Panpsychism," *The Universe of Things* (Minneapolis: University of Minnesota Press, 2014), 90.

374. Tom Sparrow, *The End of Phenomenology*, 91.

375. Peter-Paul Verbeek, "Expanding Mediation Theory," *Foundations of Science* 17 (2012): 391-395.

376. Don Ihde, *Postphenomenology: Essays in the Postmodern Context* (Evanston, Ill: Northwestern University Press, 1993), 34.

377. Ibid., 46.

378. An image of SHRDLU's graphical interface, as well as a lengthy demo dialogue, is available on the Stanford University website. "SHRDLU." http://hci.stanford.edu/winograd/shrdlu/ (accessed January 22 2018). A Windows text-only console version of SHRDLU is also accessible through this website, as is the original source code.

379. Hubert Dreyfus, "From Micro-Worlds to Knowledge Representation: AI at an Impasse," *Mind Design II*, ed. John Haugeland (Cambridge, MA: The MIT Press, 1997), 151. Originally published 1979.

380. Ibid.

381. Ibid., 145.

382. Ray Kurzweil, *How to Create a Mind* (New York: Penguin Books, 2012), 181.

383. John R. Searle, "Minds, Brains, and Program," in *The Turing Test: Verbal Behaviour as the Hallmark of Intelligence*, ed. Stuart M. Shieber (Cambridge, MA: MIT Press, 2004), 201-224.

384. Ibid.

385. Yehoshua Bar-Hillel, "The Present Status of Automatic Translation of Languages," in *Readings in Machine Translation*, eds. Sergei Nirenburg, Harold Somers, and Yorick Wilks (Cambridge, MA: The MIT Press, 2003), originally published 1960.

386. Terry Winograd, "What Does it Mean to Understand Language?" *Cognitive Science* 4 (1980).

387. Terry Winograd, "Procedures as a Representation for Data in a Computer Program for Understanding Natural Language," doctoral dissertation (Massachusetts Institute of Technology, 1971), 10.

388. Terry Winograd, "Lecture 1," *Five Lectures on Artificial Intelligence* (Springfield, VA: National Technical Information Service, 1974), 4.

389. Jason L Hutchens, "How to Pass the Turing Test by Cheating" *ACM DL Digital Library* (University of Western Australia, 1996), 6.

390. Winograd, "Lecture 1," 4.

391. Megan Garber, "Would You Want Therapy From a Computerized Psychologist?," *The Atlantic* (May 23 2014).

392. Gale M. Lucas, Jonathan Gratch, Aisha King and Louis-Phillippe Morency, "It's Only a Computer: Virtual Humans Increase Willingness to Disclose," *Computers in Human Behavior* 37 (2014): 94-100.

393. Alexis Elder, "Robot Friends for Autistic Children: Monopoly Money or Counterfeit Currency?", in *Robot Ethics 2.0: From Autonomous Cars to Artificial Intelligence*, ed. Patrick Lin, Ryan Jenkins and Keith Abney (New York: Oxford University Press, 2017), Kindle Ebook.

394. Kurzweil, *The Singularity is Near*.

395. Sherry Turkle, *Life on the Screen* (New York: Simon & Schuster Paperbacks, 1995), 130.

396. Kurzweil, *The Singularity is Near*.

397. *Bob Dylan + IBM Watson on Language* (2015), Youtube. https://www.youtube.com/watch?v=oMBUk-57FGU (accessed June 27 2017).

398. *New Era of Cognitive Computing* (2013), Youtube. https://www.youtube.com/watch?v=h22n80aT2FY (accessed June 27 2017).

399. Spike Jonze, *Her* (United States: Warner Bros. Pictures, 2013), film.

400. Winograd, "Lecture 2," in *Five Lectures on Artificial Intelligence*), 20.

401. Deb Roy, "Semiotic Schemas: a Framework for Grounding Language in Action and Perception," *Artificial Intelligence* 167 (2005), 176.

402. Terry Winograd, "Understanding Natural Language," *Cognitive Psychology* 3 (1972). Excerpts from the dialogue between human and SHRDLU can be most readily accessed on the Stanford website: "SHRDLU," http://hci.stanford.edu/winograd/shrdlu/ (accessed January 22 2018).

403. Douglas R. Hofstadter, *Gödel, Escher, Bach: An Eternal Golden Braid*. (Harmondsworth, England: Penguin Books, 1980), 630. Hofstadter's discussion of SHRDLU is extensive and insightful.

404. Winograd, "Procedures as a Representation for Data in a Computer Program for Understanding Natural Language," 443.

405. Terry Winograd and Fernando Flores, *Understanding Computers and Cognition: A New Foundation for Design* (Reading, MA: Addison Wesley Publishing Company, Inc., 1987), originally published 1986.

406. Terry Winograd and Fernando Flores, *Understanding Computers and Cognition*, quoted in Wolfgang G. Stock and Mechtild Stock, *Handbook of Information Science*, trans. Paul Becker (Berlin: De Gruyter, 2013), 56.

407. Daniel Andler, "Phenomenology in Artificial Intelligence and Cognitive Science," in *The Blackwell Companion to Phenomenology and Existentialism*, ed. H. Dreyfus and M. Wrathall (London: Blackwell, 2006), 377-393.

408. Maja Mataric, "What is a Robot? Defining Robots," *The Robotics Primer* (Cambridge, MA: The MIT Press, 2007).

409. Ibid.

410. Harman, *The Quadruple Object*, 5.

411. Michael M. Khonsari and Mehdi Amiri, "Fundamentals of Thermodynamics," *Introduction to Thermodynamics of Mechanical Fatigue* (Boca Raton: CRC Press, 2013), 11.

412. Ibid.

413. *SHRDLU*, Windows 7.

414. Verbeek, "The Acts of Artifacts," *What Things Do*, 147-172.

415. Ibid., 166.

416. Robert Rosenberger and Peter-Paul Verbeek, "A Postphenomenological Field Guide," in *Postphenomenological Investigations: Essay on Human-Technology Relations*, eds. Robert Rosenberger and Peter-Paul Verbeek (Lanham, Maryland: Lexington Books, 2015), 30.

417. Don Ihde, "Introduction: Postphenomenological Research," *Human Studies* 31, no. 1 (2008).

418. Don Ihde, *Postphenomenology and Technoscience: The Peking University Lectures*, (Albany NY: State University of New York Press, 2009), 40.

419. Don Ihde, *Bodies in Technology*, (Minneapolis: University of Minnesota Press, 2001).

420. Gert Goeminne and Erik Paredis, "Opening Up the In-Between: Ihde's Postphenomenology and Beyond," *Foundations of Science* 16, no. 2-3 (2011).

421. Don Ihde, *Technology and the Lifeworld: From Garden to Earth* (Bloomington: Indiana University Press, 1990).

422. Ihde makes a study of hearing in: Don Ihde, *Listening and Voice: Phenomenologies of Sound* (Albany: State University of New York Press, 2007).

423. Ihde, *Bodies in Technology*.

424. Ihde, *Technology and the Lifeworld*, 74.

425. Ihde, *Postphenomenology and Technoscience*.

426. Ihde, *Technology and the Lifeworld*, 73.

427. Ihde, *Postphenomenology and Technoscience*.

428. Bruno Latour, *Science in Action* (Cambridge: Harvard University Press, 1987).

429. Ihde, *Technology and the Lifeworld*, 75.

430. Crary, *Techniques of the Observer*.

431. Ken Hillis, "The Sensation of Ritual Space," *Digital Sensations: Space, Identity and Embodiment in Virtual Reality* (Minneapolis: University of Minnesota Press, 1999), 70.

432. Jonathan Crary, *Suspensions of Perception: Attention, Spectacle and Modern Culture* (Cambridge, MA: MIT Press, 1999).

433. Ihde, *Technology and the Lifeworld*, 81.

434. Ibid., 85.

435. B. Koribalski, K. Jones, M. Elmouttie and R. Haynes, "Neutral Hydrogen Gas in the Circinus Galaxy," *Australia Telescope Outreach and Education, CSIRO*. http://outreach.atnf.csiro.au/images/astronomical/circinus.html (accessed June 27 2017).

436. Ihde, *Technology and the Lifeworld*, 98.

437. Ihde, *Bodies in Technology*, 81.

438. *Technology and the Lifeworld.*, 109.

439. Ibid., 112.

440. Ibid., 145.

441. Robert Rosenberger, "Multistability and the Agency of Mundane Artifacts: from Speed Bumps to Subway Benches," *Human Studies* 37 (2014), 378.

442. Evan Selinger, "Normative Judgement and Technoscience: Nudging Ihde, Again," *Techné* 12, no. 2 (2008): 120-125.

443. *Halo 4*, developed by 343 Industries (Microsoft Studios: 2012), Xbox 360.

444. Martin Gibbs, Marcus Carter, Michael Arnold, "Avatars, Characters, Players and Users: Multiple Identities at/in Play," paper presented at *OZCHI'12*, Melbourne, Australia, November 26-30 2012.

445. *Operation: Pedopriest,* developed by Molleindustria (2007), PC.

446. Ian Bogost, "The Rhetoric of Video Games," in *The Ecology of Games: Connecting Youth, Games, and Learning*, ed. Katie Salen, The John D. and Catherine T. MacArthur Foundation Series on Digital Media and Learning, (Cambridge, MA: The MIT Press, 2008), 117–140.

447. Marvin Minsky, *The Society of Mind* (London: Picador, 1988).

448. Ibid., "1.4: The world of blocks."

449. Ibid., 21.

450. Winograd, "Understanding Natural Language," 5.

451. Ibid.,4-5.
452. Hofstadter, *Gödel, Escher, Bach: An Eternal Golden Braid*, 628.
453. John Law "Notes on the Theory of the Actor-Network: Ordering, Strategy and Heterogeneity," *Systems Practice* 5 (1992), 385.
454. Mark C. Marino, "The Racial Formation of Chatbots," *CLCWeb: Comparative Literature and Culture* 16, no. 5, article 13 (2014).
455. Cathrine Hasse, "Artefacts That Talk: Mediating Technologies as Multistable Signs and Tools," *Subjectivity* 6, no. 1 (2013): 79-100.
456. Ibid., 87.
457. Robert Rosenberger, "The Sudden Experience of the Computer," *AI & Society* 24 (2009): 173-180.
458. Ibid., 174.
459. Ibid., 145.
460. P. Brey, "Technology and Embodiment in Ihde and Merleau-Ponty," in *Metaphysics, Epistemology, and Technology. Research in Philosophy and Technology* volume 19, ed. C. Mitcham (London: Elsevier, 2000), 9.
461. Yoni van den Eede, "The Mediumness of World: A Love Triangle of Postphenomenology, Media Ecology, and Object-Oriented Philosophy," in *Postphenomenology and Media: Essays on Human-Media-World Relations*, ed. Yoni van den Eede, Stacey O'Neal Irwin, and Galit Wellner (Lanham: Lexington Books, 2017), 229-250.
462. Ibid., 246
463. Ibid., 240
464. Ibid., 234
465. Ibid., 238
466. Ihde, *Technology and the Lifeworld*, 29.
467. Verbeek, *What Things Do*, 122.
468. Anette Forss, "Cells and the (Imaginary) Patient: the Multistable Practitioner-Technology-Cell Interface in the Cytology Laboratory," *Medicine, Health Care and Philosophy* 15, no. 3 (2012), 298.
469. *Advance Female Android Aiko AI Robot Fembot* (2007), Youtube. https://www.youtube.com/watch?v=iCR2PFrLkwA (accessed June 27 2017).
470. Some volumes published on this subject include: John Danaher and Neil McArthur, ed. *Robot Sex: Social and Ethical Implications* (Cambridge, MA: MIT Press, 2017). David Levy, *Love & Sex with Robots: The Evolution of Human-Robot Relationships* (London: Duckworth Overlook, 2009), originally published 2007. And Kate Devlin, *Turned On: Science, Sex and Robots* (London: Bloomsbury Sigma, 2018).
471. Bryant, *Onto-Cartography*, 26.
472. Opinions vary on whether sex robots currently exist. John Danaher states that two products discussed in this chapter (Roxxxy and Harmony) do really exist, despite

being "crude". John Danaher, "Should We Be Thinking about Robot Sex?", in *Robot Sex*, 6.

473. Craig Gillespie, *Lars and the Real Girl*, (United States and Canada: Metro-Goldwyn-Meyer) 2007, film.

474. "Realbotix," *Realbotix*. https://realbotix.com/Index2# (accessed May 9 2018).

475. Donna Haraway, "Situated Knowledges: the Science Question in Feminism and the Privilege of Partial Perspective," *Feminist Studies* 14, no. 3 (Autumn 1988), 579.

476. Ian Bogost, "Objects and Videogames: Why I am Interested in Both," *Ian Bogost* (blog) (June 19 2010). http://bogost.com/writing/blog/objects_and_videogames/ (accessed June 27 2017).

477. Morton, *Hyperobjects*.

478. Graham Harman, *Immaterialism*, ed. Laurent de Sutter, Theory Redux (Cambridge, UK: Polity Press, 2016).

479. Ibid., 29.

480. Dorothy Howard, "Loving Machines: a De-Anthropocentric View of Intimacy," *Arachne*, no. 00 (2015).

481. Braidotti, "Post-Humanism: Life Beyond the Self," *The Posthuman*.

482. Ibid., 15.

483. Ibid., 29.

484. Masahiro Mori, *The Buddha in the Robot*, trans. Charles S. Terry (Tokyo: Kosei Publishing Co., 1981).

485. Ibid., 21.

486. Eun Hwa Jung, T. Franklin Waddell and S Shyam Sundar, "Feminizing Robots: User Responses to Gender Cues on Robot Body and Screen," *Conference on Human Factors in Computing Systems Proceedings* (2016): 3107-13.

487. "ASIMO Frequently Asked Questions," *Honda*. http://Asimo.honda.com/news/ (accessed June 22 2016).

488. Roger Andre Søraa conjectures that this may derive from a difference between the English and Japanese languages. In Japanese pronouns are used less often and gendering is more evident in the addition of suffixes to names (such as "-san" and other suffixes that may indicate gender or endearment). Roger Andre Søraa, "Mechanical Genders: How Do Humans Gender Robots?" *Gender, Technology and Development* 21, no. 1-2 (2017): 99-115.

489. Jennifer Robertson, "Gendering Robots: Posthuman Traditionalism in Japan," in *Recreating Japanese Men*, ed. Sabine Frühstück and Anne Walthall, (Berkeley: University of California Press, 2011), 320.

490. "FT [FEMALE TYPE]," *Robo Garage* (2009). http://robo-garage.com (accessed June 28 2017).

491. Judith Butler, "Critically Queer," *GLQ* 1 (1993), 18.

492. Judith Butler, *Gender Trouble: Feminism and the Subversion of Identity* (New York: Routledge, 1999), originally published 1990, 25, quoted in Sara Salih, "On Judith Butler and Performativity," in *Sexualities and Communication in Everyday Life: A*

Reader, eds. Karen E. Lovaas and Mercilee M. Jenkins (Thousand Oaks, CA: SAGE Publications, 2007), 56.

493. Erico Guizzo, "Hiroshi Ishiguro: the Man Who Made a Copy of Himself," *IEEE Spectrum* (2010).

494. Hanson Robotics, "Bina48," (2017). http://www.hansonrobotics.com/robot/bina48/ (accessed May 20 2018).

495. A video of BINA48 and Bina Rothblatt's meeting can be seen here: The LifeNaut Project, *Bina 48 meets Bina Rothblatt* (November 27 2014), Youtube. https://www.youtube.com/watch?v=KYshJRYCArE (accessed May 9 2018).

496. *Harmony AI*, developed by Realbotix (2017), Android (accessed April 20 2018).

497. "Truecompanion.com," *True Companion* (2016). http://www.truecompanion.com/home.html (accessed May 25 2016).

498. David Levy, "Roxxxy the "Sex Robot" - Real or Fake," *Lovotics* 1 (2013).

499. In an interview in January 2018, Roxxxy informed viewers that she would not be available to the public for a while. See TruecompanionLLC, *Roxxxy Sex Robot Meets Dr Oz* (2018), Youtube. https://www.youtube.com/watch?v=54Ojv33Nm9Y (accessed June 18 2018).
As of May 2019 I have been unable to find a single user's comments or report suggesting that they have received or used a Roxxxy. John Danaher also comes to this conclusion but says that he remains "agnostic" about whether she will ever be available. See John Danaher, "Should We Be Thinking about Robot Sex?", in *Robot Sex*, 7.

500. "Truecompanion.com," *True Companion*.

501. Ridley Scott, *Blade Runner* (United States: Warner Bros., 1982), , film.

502. Gregory Jerome Hampton, "The True Cult of Humanhood: Displacing Repressed Sexuality Onto Mechanical Bodies," *Imagining Slaves and Robots in Literature, Film, and Popular Culture* (Lanham: Lexington Books, 2015).

503. Scott, *Blade Runner*.

504. Hampton, *Imagining Slaves and Robots in Literature*, 31.

505. Ovid, "Metamorphoses," in *The Ovid Collection*, trans. A.S. Kline (2010). http://ovid.lib.virginia.edu/trans/Ovhome.htm (accessed June 27 2017). See also Kenneth Gross's analysis in Part 2 of *The Dream of the Moving Statue* (University Park, Pennsylvania: The Pennsylvania State University Press, 2006), especially pages 72-79.

506. George Cukor, *My Fair Lady* (United States: Warner Bros., 1964), film.

507. Ovid, "Metamorphoses," "Book X: 243-297 Orpheus Sings: Pygmalion and the Statue."

508. William Morris, "Pygmalion and the Image," *Sacred Texts*, originally published 1868. http://www.sacred-texts.com/neu/morris/ep1/ep121.htm (accessed June 27 2017).

509. A. Scobie and J. Taylor, "Perversions Ancient and Modern: I. Agalmatophilia, the Statue Syndrome," *Journal of the History of the Behavioral Sciences* 11, no. 1 (1975): 49-54.

510. Julie Wosk, *My Fair Ladies: Female Robots, Androids and Other Artificial Eves* (New Jersey: Rutgers University Press, 2015), 39.
The automaton is housed at the *Musée des arts et métiers* in Paris, and photos of this beautiful object can be found on their website.

511. "RealDoll," *RealDoll* (2015), https://secure.realdoll.com/ (accessed February 18 2016).

512. Julie Beck, "Married to a Doll: Why One Man Advocates Synthetic Love," The Atlantic (September 6 2013). http://www.theatlantic.com (accessed January 20 2018).

513. Devlin, *Turned On*, 'Chapter Eight: Utopia/Dystopia'.

514. Stacy Leigh, "average americans (that happen to be sex dolls)" *stacy leigh*, www.stacythearist.com (accessed May 10 2019).

515. Donald W. Black and Jon E. Grant, "Sexual Dysfunctions, Gender Dysphoria, and Paraphilic Disorders," *DSM-5 Guidebook* (Washington DC: American Psychiatric Publishing, 2014), 286. There is also a small subculture of individuals who form romantic and/or sexual relationships with other kinds of nonhumans, such as cars, furniture and public monuments. This subculture may be accessed through *Objectùm-Sexuality Internationale*, http://www.objectum-sexuality.org/ (accessed June 27 2017).

516. American Psychiatric Association, "Paraphilic disorders," *Diagnostic and Statistical Manual of Mental Disorders* fifth edition, DSM Library. http://dsm.psychiatryonline.org (accessed August 22 2017).

517. Beck, "Married to a Doll."

518. Judith A. Markowitz, "Cultural Icons," in *Robots that Talk and Listen: Technology and Social Impact*, ed. Judith A. Markowitz (Berlin: De Gruyter, 2015), 41.

519. Allison deFren, *asfr* (2001), Youtube. https://www.youtube.com/watch?v=hfHs1xQTZ2E (accessed June 27 2017).

520. Melanie Ehrenkranz, "Futuristic Fembot "Jia Jia" is One of the Most Sexist Robot Creations Yet," *Tech.Mic* (2016). https://mic.com (accessed June 27 2017).

521. "[Independent] Chinese Researchers Create Jia Jia - a Super-Lifelike 'Robot Goddess'," *University of Science and Technology of China* (April 19 2016). http://en.ustc.edu.cn (accessed June 27 2017).

522. Le Trung, "When Science Meets Beauty," *Project Aiko* (2007-2013). http://www.projectaiko.com/index.html (accessed June 29 2017).

523. Le Trung, e-mail message to author (March 4 2016).

524. Elizabeth Monk-Turner, "Gender and Market Opportunity in Japan," *National Journal of Sociology* 11 (1997).

525. *Advance Female Android Aiko AI Robot Fembot*.

526. Trung, "When Science Meets Beauty."

527. Ibid.

528. Female avatars are increasingly a part of the infrastructure of the home, and part of the reason may be to replace or simulate the presence of a "wife". Jenny Kennedy, "Addressing the Wife Drought: Automated Assistants in the Smart

Home," presentation at *At Home with Digital Media*, Queensland University of Technology, November 2-3 2018.

529. Behar, "The Other Woman."

530. Roger Sherman Loomis, *Celtic Myth and Arthurian Romance* (Chicago: Academy Chicago Publishers, 2005), 17.

531. Bryan Forbes, *The Stepford Wives* (Columbia Pictures, 1975), film.

532. Scott, *Blade Runner*.

533. Minsoo Kang, "Building the Sex Machine: The Subversive Potential of the Female Robot," *Intertexts* 9(1) (2005): 5-22.

534. Gaby Wood, "Chapter 3: Journey to the Perfect Woman," in *Edison's Eve: A Magical History of the Quest for Mechanical Life* (New York: Alfred A. Knopf, 2002).

535. Ibid., 122.

536. David Firestone, "While Barbie Talks Tough, G.I. Joe Goes Shopping," *The New York Times* (December 31 1993). http://www.nytimes.com (accessed January 21 2018).

537. Kim Toffoletti, "Barbie: A Posthuman Prototype," *Cyborgs and Barbie Dolls* (London: I.B. Tauris, 2007), 63.

538. Wosk, *My Fair Ladies*, 39.

539. Ibid., 41.

540. Ibid., 140.

541. Fritz Lang, *Metropolis* (Germany: UFA, 1927), film.

542. Takayuki Tatsumi, *Full Metal Apache: Transactions Between Cyberpunk Japan and Avant-Pop America* (Durham: Duke University Press, 2006), 213.

543. Gwyneth Jones, *Divine Endurance* (London: George Allen & Unwin, 1984).

544. *Ibid.*, 11.

545. Ibid., 140.

546. Wendell Wallach and Colin Allen, "Why Machine Morality?" *Moral Machines: Teaching Robots Right From Wrong* (Oxford: Oxford University Press, 2009).

547. Tessa Leach, "Who Is Their Person? Sex Robot and Change." *Queer-Feminist Science and Technology Studies Forum* 3 (December 2018): 25-39.

548. Harman, *Guerrilla Metaphysics*, 137.

549. Alex Garland, *Ex Machina* (United Kingdom and the United States: Universal Studios: 2014), film.

550. "NASA Awards Two Robots to University Groups For R&D Upgrades," *NASA* (2015). https://www.nasa.gov (accessed June 28 2017).

551. Jorge Luis Borges, *The Book of Imaginary Beings*, trans. Norman Thomas di Giovanni (Harmondsworth, England: Penguin Books, 1984), originally published 1967, 151.

552. Jennifer Robertson, "Gendering Humanoid Robots: Robo-Sexism in Japan," *Body & Society* 16, no. 2 (2010): 1-36.

553. A recent study suggests that Americans are more accepting of anthropomorphic qualities in robots than are Japanese people, while acknowledging that the perception of humanity in robots must be measured differently between cultures. See: Hiroko Kamide and Tatsuo Arai, "Perceived Comfortableness of Anthropomorphized Robots in U.S. and Japan," *International Journal of Social Robotics* 9 (September 2017): 537-543.

554. Martin Ford, *Rise of the Robots: Technology and the Threat of Mass Unemployment* (London: Oneworld Publications, 2016).

555. Frederik L. Schodt, *Inside the Robot Kingdom: Japan, Mechatronics and the Coming Robotopia* (Tokyo: Kodansha International Ltd., 1988), 14.
Karl MacDorman, Sandosh Vasudevan and Chin-Chang Ho, "Does Japan Really Have Robot Mania? Comparing Attitudes by Implicit and Explicit Measures," *AI & Society* 23, no. 4 (2009): 485-510.

556. Robots may serve in numerous aspects of Japanese life. There are reports of robots being given their own funerals. See Miwa Suzuki, "In Japan, Aibo Robots get their own Funeral," *The Japan Times* (May 1 2018). https://www.japantimes.co.jp (accessed June 18 2018). Some robots, such as the robot Pepper, have also served as monks and priests in Japan. See Leon Siciliano and Reuters, "A Japanese Company just Unveiled a Robot Priest that will read Scriptures at Buddhist Funerals," *Business Insider* (August 24 2017). http://www.businessinsider.com (accessed June 18 2018).

557. Simons, *Robots*, 29.

558. Justin Lewis Bernstein, "The Warm Welcome of the Japanese Cyborg," *Journal of Asia Pacific Studies* 3, no.1 (2013), 118.

559. Judith Butler, *Gender Trouble: Feminism and the Subversion of Identity*, ed. Lind J. Nicholson, Thinking Gender (New York: Routledge, 2006), originally published 1990.

560. Lily Frank and Sven Nyholm, "Robot Sex and Consent: is Consent to Sex between a Robot and a Human Conceivable, Possible, and Desirable?" *Artificial Intelligence and Law* 25, no. 3 (2017): 305-323.

561. David Levy, "The Mental Leap to Sex with Robots," in *Love & Sex with Robots*: 274-302.

562. Noel Sharkey, Aimee van Wynsberghe, Scott Robbins and Eleanor Hancock, *Our Sexual Future With Robots* (The Hague: Foundation for Responsible Robotics, 2017), PDF. And Litska Strikwerda, "Legal and Moral Implications of Child Sex Robots," in *Robot Sex*, 133-152.

563. Rachael Revesz, "Paedophiles 'Could be Prescribed Child Sex Dolls' to Prevent Real Attacks, Says Therapist," *Independent* (August 2 2017). http://www.independent.co.uk (accessed January 8 2018).

564. Providing sex doll therapy to treat mental illness is reminiscent of the use of vibrators for the treatment of hysteria in women during the nineteenth century. Doctors resisted the confusion of clinical therapies with the consumer models that became popular after 1900. See Rachel P. Maines, *The Technology of Orgasm* (Baltimore: The Johns Hopkins University Press, 1999), 93-95.

565. *Campaign Against Sex Robots* (2016). https://campaignagainstsexrobots.wordpress.com (accessed June 22 2016).

566. Kathleen Richardson, "Urgent – Why We Must Campaign to Ban Sex Dolls & Sex Robots." *Campaign Against Sex Robots* (October 3 2017). https://campaignagainstsexrobots.org/2017/10/03/urgent-why-we-must-campaign-to-ban-sex-dolls-sex-robots/ (accessed May 26 2018).

567. John Danaher, Brian Earp, and Anders Sandberg, "Should we Campaign Against Sex Robots?" in *Robot Sex*, 53.

568. Julia Long, *Anti-Porn: The Resurgence of Anti-Pornography Feminism* (London: Zed Books, 2012).

569. "Ethics of Robotics," *Campaign Against Sex Robots* (2016). https://campaignagainstsexrobots.wordpress.com/ethics-of-robots/ (accessed June 23 2016).

570. Steve Petersen discusses this question but primarily with reference to future or imagined robots in "Is it Good for Them Too? Ethical Concern for the Sexbots," in *Robot Sex*, 155-172.

571. Kate Darling, "Extending Legal Rights to Social Robots," *IEEE Spectrum* (2012).

572. John Basl, "Machines as Moral Patients We Shouldn't Care About (Yet): the Interests and Welfare of Current Machines," *Philosophy & Technology* 27, no. 1 (March 2014): 79-96.
For a discussion of whether consciousness is necessary for "moral agency" see: Himma, "Artificial Agency, Consciousness."

573. Darling, "Extending Legal Rights to Social Robots."

574. Kathleen Richardson, "Sex Robot Matters," *IEEE Technology and Society Magazine* (June 2016).

575. For example in this sensationalist article: Siobhan McFadyen, "Shocking Tiny SEX ROBOT Which Looks Like Schoolgirl is on Sale for £770 and Comes Delivered in 'Coffin'," *Mirror* (March 16 2016). http://www.mirror.co.uk (accessed January 21 2018).

576. Latour, *Irreductions*, 168.

577. Robert Sparrow tackles this issue in "Robots, Rape, and Representation," *International Journal of Social Robotics* 9 (September 2017): 465-477.

578. Bruno Latour, *We Have Never Been Modern*, trans. Catherine Porter (Hertfordshire: Harvester Wheatsheaf, 1993).

579. Latour, "On the Difficulty of Being an ANT," *Reassembling the Social*, 148.

580. Bruno Latour, *Pandora's Hope: Essays on the Reality of Science Studies* (Cambridge, MA: Harvard University Press, 1999), 180.

581. Langdon Winner has criticised social constructivism of technology as a whole including authors such as Latour for this: Langdon Winner, "Upon Opening the Black Box and Finding it Empty: Social Constructivism and the Philosophy of Technology," *Science, Technology and Human Values* 18, no. 3 (Summer 1993): 362-378.

582. "Lower" and "raise" are in scare quotes here because it suggests a hierarchy of objects, which is not the intention.

583. Harman, "On Vicarious Causation," 215.

584. Mason, "Loving the Technological Undead: Cyborg Sex and Necrophilia in Richard Calder's Dead Trilogy," in *The Body's Perilous Pleasures*, ed. Michelle Aaron, (Edinburgh: Edinburgh University Press, 1999) 108-125.

585. Ian G. R. Shaw and Katherine Meehan, "Force-Full: Power, Politics and Object-Oriented Philosophy," *Area* 45, no. 2 (2013), 217.

586. N. Katherine Hayles, "Speculative Aesthetics and Object-Oriented Inquiry (OOI)," *Speculations: A Journal of Speculative Realism* vol. V (2014): 158-179.

587. Ibid., 160.

588. Ibid., 164.

589. Ibid., 168.

590. Ibid., 177.

591. Ibid., 177.

592. Susan R. Fussell, Sara Kiesler, Leslie D. Setlock, and Victoria Yew, "How people anthropomorphize robots," paper presented at *HRI*, Amsterdam, March 2008.

593. *Jordan Wolfson: (Female Figure) 2014* (2014), Youtube. https://www.youtube.com/watch?v=mVTDypgmFCM (accessed June 29 2017).

594. "Beyond beyond the valley of the dolls," *scanlines* (2015). http://scanlines.net (accessed June 29 2017).

595. Malin, "Communicating with Objects."

596. Michael O'Rourke, "Girls welcome!!!" in *After the "Speculative Turn": Realism, Philosophy, and Feminism*, eds. Katerina Kolozova and Eileen A. Joy (Earth: punctum books, 2016), 159-197.

597. Behar, "The Other Woman."

598. Kate Darling, "'Who's Johnny?'"

Bibliography

Aarseth, Espen J. *Cybertext: Perspectives on Ergodic Literature*. Baltimore, Maryland: John Hopkins University Press, 1997. Google Ebook.

"Accessibility, the Xbox One and Kinect." Xbox. 2014. http://support.xbox.com/en-US/xbox-one/system/accessibility (accessed April 30 2014).

Ackerman, Spencer, and James Ball. "Optic Nerve: Millions of Yahoo Webcam Images Intercepted by GCHQ." *The Guardian* (February 28 2014). https://www.theguardian.com/ (accessed January 20 2018).

Advance Female Android Aiko AI robot fembot. 2007. Youtube. https://www.youtube.com/watch?v=iCR2PFrLkwA (accessed June 27 2017).

Albert, Brian. "Kinect Sports Rivals Review: a Swing and a Hit." *IGN Australia* (April 7 2014). http://au.ign.com/ (accessed May 29 2014).

Allison, Fraser, Marcus Carter and Martin Gibbs. "Word play: a history of voice interaction in digital games." *Games and Culture* (2017): 1-23.

American Psychiatric Association. "Paraphilic Disorders." In *Diagnostic and Statistical Manual of Mental Disorders*. Fifth edition. DSM Library. http://dsm.psychiatryonline.org (accessed August 22 2017).

Andler, Daniel. "Phenomenology in Artificial Intelligence and Cognitive Science." In *The Blackwell Companion to Phenomenology and Existentialism*. Edited by H. Dreyfus and M. Wrathall, 377-393. London: Blackwell, 2006.

Angus, Ian. "When did the Anthropocene Begin… and Why Does it Matter?" *Monthly Review: An Independent Socialist Magazine* 67, no. 4 (September 2015): 1-11.

Apperley, Thomas. "The Body of the Gamer: Game Art and Gestural Excess." *Digital Creativity* 24, no. 2 (2013): 145-156.

Apperley, Tom and Nicole Heber. "Capitalising on Emotions: Digital Pets and the Natural User Interface." In *Game Love: Essays on Play and Affection*. Edited by J. Evenold and E. MacCallum-Stewart, 149-164. Jefferson, NC: McFarland and Company, 2015.

Arnold, Michael. "Da Vinci and Me: a Case Study in the Postphenomenology of Robotic Surgery." Paper presented at *Phenomenology, Imagination and Virtual Reality*, Husserl Archives Institute of Philosophy KU Leuven, May 30 2017.

"ASIMO Frequently Asked Questions." Honda. http://Asimo.honda.com/news/ (accessed June 22 2016).

Bar-Hillel, Yehoshua. "The Present Status of Automatic Translation of Languages." In *Readings in Machine Translation*. Edited by Sergei Nirenburg, Harold Somers and Yorick Wilks. Cambridge, MA: The MIT Press, 2003. Originally published 1960.

Basl, John. "Machines as Moral Patients We Shouldn't Care About (Yet): the Interests and Welfare of Current Machines." *Philosophy & Technology* 27, no. 1 (March 2014): 79-96.

Beck, Julie. "Married to a Doll: Why One Man Advocates Synthetic Love." *The Atlantic* (September 6 2013). http://www.theatlantic.com (accessed January 20 2018).

Behar, Katherine. "An Introduction to OOF." In *Object-Oriented Feminism*. Edited by Katherine Behar, 1-36. Minneapolis: University of Minnesota Press, 2016.

———. "The Other Woman." In *After the "Speculative Turn": Realism, Philosophy, and Feminism*. Edited by Katerina Kolozova and Eileen A. Joy, 27-38. Earth: punctum books, 2016.

Benhur Oral, Sevket. "Weird Reality, Aesthetics, and Vitality in Education." *Studies in Philosophy and Education* 34, no. 5 (2014): 459-474.

Bennett, Jane. "Systems and Things: A Response to Graham Harman and Timothy Morton," *New Literary History* 43, no.2 (Spring 2012): 225-233.

———. *Vibrant Matter: a Political Ecology of Things*. Durham: Duke University Press, 2010. Kindle Ebook.

Bernstein, Justin Lewis. "The Warm Welcome of the Japanese Cyborg." *Journal of Asia Pacific Studies* 3, no. 1 (2013): 110-154.

"Beyond Beyond the Valley of the Dolls." *scanlines* (2015). http://scanlines.net/ (accessed June 29 2017).

Black, Donald W., and Jon E. Grant. *DSM-5 Guidebook*. Washington DC: American Psychiatric Publishing, 2014.

Bob Dylan + IBM Watson on Language. 2015. Youtube. https://www.youtube.com/watch?v=oMBUk-57FGU (accessed June 27 2017).

Boden, Margaret. "Grey Walter's anticipatory tortoises." *The Rutherford Journal* 2 (2006-7).

Bogost, Ian. "The Aesthetics of Philosophical Carpentry." In *The Nonhuman Turn*. Edited by Richard Grusin, 81-100. Minneapolis: University of Minnesota Press, 2015.

———. *Alien Phenomenology, or What It's Like to Be a Thing*. Posthumanities. Minneapolis: University of Minnesota Press, 2012.

———. "Inhuman." In *Inhuman Nature*. Edited by Jeffrey Jerome Cohen. Washington DC: Oliphaunt Books, 2014.

———. "Objects and Videogames: Why I am Interested in Both." Ian Bogost (blog) (June 19 2010). http://bogost.com/writing/blog/objects_and_videogames/ (accessed June 27 2017).

———. "The Rhetoric of Video Games." In *The Ecology of Games: Connecting Youth, Games, and Learning*. Edited by Katie Salen, 117-140. The John D. and Catherine T. MacArthur Foundation Series on Digital Media and Learning. Cambridge, MA: The MIT Press, 2008.

———. "Videogames are a mess." Ian Bogost (blog) (September 3 2009). http://bogost.com/writing/videogames_are_a_mess/ (accessed June 26 2017).

Borges, Jorge Luis. *The Book of Imaginary Beings*. Translated by Norman Thomas di Giovanni. Harmondsworth, England: Penguin Books, 1984. Originally published 1967.

Braidotti, Rosi. *The Posthuman*. Cambridge, UK: Polity Press, 2013.

Brentari, Carlo. *Jakob von Uexküll: The Discovery of the Umwelt between Biosemiotics and Theoretical Biology*. Biosemiotics. Dordrecht: Springer, 2015.

Brey, P. "Technology and Embodiment in Ihde and Merleau-Ponty." In *Metaphysics, Epistemology, and Technology. Research in Philosophy and Technology* vol. 19. Edited by C. Mitcham, 45-58. London: Elsevier, 2000.

Brier, Søren. "Cybersemiotics and Umweltehre." *Semiotica* 134, no. 1-4 (2001): 779-814.

———. *Cybersemiotics: Why Information Is Not Enough!* Toronto: University of Toronto Press, 2008.

Brown, Nathan. "The Nadir of OOO: from Graham Harman's Tool-Being to Timothy Morton's Realist Magic: Objects, Ontology, Causality." *Parrhesia*, no. 17 (2013): 62-71.

Bryant, Levi. The Democracy of Objects. *New Metaphysics*. University of Michigan Library: Open Humanities Press, 2012. PDF Ebook.

———. Larval Subjects (blog). http://larvalsubjects.wordpress.com

———. *Onto-Cartography: An Ontology of Machines and Media*. Edinburgh: Edinburgh University Press, 2014.

Bryant, Levi, Nick Srnicek, and Graham Harman. *The Speculative Turn: Continental Materialism and Realism*. Anamnesis. Melbourne: re.press, 2011.

Butler, Judith. "Critically queer." *GLQ* 1 (1993): 17-32.

———. *Gender Trouble: Feminism and the Subversion of Identity*. Edited by Lind J. Nicholson. Thinking Gender. New York: Routledge, 2006. Originally published 1990.

Buyya, Rajkumar and Amir Vahid Dastjerdi, eds. *Internet of Things: Principles and Paradigms*. Amsterdam: Elsevier, 2016.

Callon, Michel. "Some Elements of a Sociology of Translation: Domestication of the Scallops and the Fishermen of St Brieuc Bay." In *Power, Action and Belief: A New Sociology of Knowledge*. Edited by John Law. London: Routledge & Kegan Paul, 1986.

"Campaign Against Sex Robots." Campaign Against Sex Robots (2016). https://campaignagainstsexrobots.wordpress.com (accessed June 22 2016).

Carter, Chris. "Impressions: Kinect Sports Rivals Preseason." *Destructoid* (November 28 2013). http://www.destructoid.com (accessed June 27 2017).

Chalmers, David J. *The Conscious Mind*. New York: Oxford University Press, 1996.

Chisholm, Roderick M. "Identity through time." *Person and Object: A Metaphysical Study*. Milton Park, UK: Taylor & Francis, 2013, 88-112. Originally published 1976.

Coeckelbergh, Mark. "Humans, Animals, and Robots: a Phenomenological Approach to Human-Robot Relations." *International Journal of Social Robotics* 3 (2011): 197-204.

Cohen, Jeffrey Jerome. *Stone: An Ecology of the Inhuman*. Minneapolis: University of Minnesota Press, 2015.

Cohen, Jeffrey Jerome, and Julian Yates, eds. *Object Oriented Environs*. Earth: punctum books, 2016.

Cohen, Jeffrey Jerome, and Linda T. Elkins-Tanton. *Earth*. Object Lessons. New York: Bloomsbury, 2017.

Commander Cherry's Puzzled Adventure. Developed and published by Grande Games (2015). Xbox One.

Crary, Jonathan. *Suspensions of Perception: Attention, Spectacle and Modern Culture*. Cambridge, MA: MIT Press, 1999.

———. *Techniques of the Observer: On Vision and Modernity in the Nineteenth Century*. Cambridge, MA: MIT Press, 1990.

"Creating a Champion in Kinect Sports Rivals." *Xbox Wire* (2014). http://news.xbox.com (accessed May 29 2014).

Crichton, Michael. *Westworld*. United States: MGM, 1973. Film.

Danaher, John. "Should We Be Thinking about Robot Sex?" In *Robot Sex: Social and Ethical Implications*. Edited by John Danaher and Neil McArthur, 3-14. Cambridge, MA: MIT Press, 2017.

Danaher, John, Brian Earp, and Anders Sandberg. "Should we Campaign against Sex Robots?" in *Robot Sex: Social and Ethical Implications*. Edited by John Danaher and Neil McArthur, 47-72. Cambridge, MA: MIT Press, 2017.

Darling, Kate. "Extending Legal Rights to Social Robots." *IEEE Spectrum* (2012).

———. "'Who's Johnny?' Anthropomorphic Framing in Human-Robot Interaction, Integration, and Policy." In *Robot Ethics 2.0: From Autonomous Cars to Artificial Intelligence*. Edited by Patrick Lin, Ryan Jenkins, and Keith Abney. New York: Oxford University Press, 2017. Kindle Ebook.

deFren, Allison. *asfr*. 2001. Youtube. https://www.youtube.com/watch?v=hfHs1xQTZ2E (accessed June 27 2017).

Devlin, Kate. *Turned On: Science, Sex and Robots* (London: Bloomsbury Sigma, 2018).

Chapter 1: Hey! Let's play games!" *Replay: The History of Videogames*. Lewes, UK: Yellow Ant, 2010.

Downs, John, Frank Vetere, and Steve Howard. "Paraplay: Exploring Playfulness Around Physical Console Gaming." In *Human-Computer Interaction – INTERACT*. Edited by P. Kotzé, G. Marsden, J. Wessen and M. Winckler, 682-699. Berlin: Springer, 2013.

Dreyfus, Hubert. "From Micro-worlds to Knowledge Representation: AI at an Impasse." *Mind Design II*. Edited by John Haugeland. Cambridge, MA: The MIT Press, 1997: 143-182. Originally published 1979.

Duffy, Brian R. "Anthropomorphism and robotics." *The Society for the Study of Artificial Intelligence and the Simulation of Behaviour – AISB* (2002).

Du Toit, Cornel. "Panpsychism, Pan-Consciousness and the Non-Human Turn: Rethinking Being as Conscious Matter." *Theological Studies* 72, no. 4 (2016): 1-11.

Dyer-Witheford, Nick, and Greig de Peuter. *Games of Empire: Global Capitalism and Video Games*. Edited by Mark Poster, Samuel Weber and Katherine Hayles. Electronics Mediations. Minneapolis: University of Minnesota Press, 2009.

E3 2009: Project Natal Xbox 360 Announcement. 2009. Youtube. https://www.youtube.com/watch?v=p2qlHoxPioM (accessed June 24 2017).

E3 2010 Microsoft - Kinect (Project Natal) Part 1. 2010. Youtube. https://www.youtube.com/watch?v=bxp43T2JP18 (accessed June 24 2017).

Ehrenkranz, Melanie. "Futuristic Fembot 'Jia Jia' is One of the Most Sexist Robot Creations Yet." *Tech.Mic* (2016).
https://mic.com (accessed June 27 2017).

Elder, Alexis. "Robot Friends for Autistic Children: Monopoly Money or Counterfeit Currency?" In *Robot Ethics 2.0: From Autonomous Cars to Artificial Intelligence*. Edited by Patrick Lin, Ryan Jenkins, and Keith Abney. New York: Oxford University Press, 2017. Kindle Ebook.

Emmeche, Claus. "Does a Robot Have an Umwelt? Reflections on the Qualitative Biosemiotics in Jakob Von Uexküll." *Semiotica* 134, no. 1-4 (2001): 653-93.

Esteban, Pablo G. et al. "How to Build a Supervised Autonomous System for Robot-Enhanced Therapy for Children with Autism Spectrum Disorder." *Paladyn, Journal of Behavioral Robotics* 8 (2017): 18-38.

Fable: The Journey. Developed by Lionhead Studios. Published by Microsoft Studios, 2012. Xbox 360.

Ferris, Timothy. "Timothy Ferris on Voyagers' Never-Ending Journey." *Smithsonian Magazine* (May 2012).
http://www.smithsonianmag.com (accessed June 7 2017).

Firestone, David. "While Barbie Talks Tough, G.I. Joe Goes Shopping." *The New York Times* (December 31 1993).
http://www.nytimes.com (accessed January 21 2018).

Flynn, Bernadette. "Geography of the Digital Hearth." *Information, Communication and Society* 6, no. 4 (2003): 551-76.

Forbes, Bryan. *The Stepford Wives*. Columbia Pictures. 1975. Film.

Ford, Martin. *Rise of the Robots: Technology and the Threat of Mass Unemployment*. London: Oneworld Publications, 2016.

Forss, Anette. "Cells and the (Imaginary) Patient: the Multistable Practitioner-Technology-Cell Interface in the Cytology Laboratory." *Medicine, Health Care and Philosophy* 15, no. 3 (2012): 295-308.

Forza Motorsport 5. Developed by Turn 10 Studios. Published by Microsoft Studios. 2013. Xbox One.

Foster, Jay. "Ontologies without Metaphysics: Latour, Harman and the Philosophy of Things." *Analecta Hermeneutica* 3 (2011): 1-26.

Frank, Lily, and Sven Nyholm. "Robot sex and consent: is consent to sex between a robot and a human conceivable, possible, and desirable?" *Artificial Intelligence and Law* 25, no. 3 (2017): 305-323.

Freeman, Will. "Kinect Sports Rivals Review - Going Through the Motions." *The Guardian* (April 11 2014).
http://www.theguardian.com (accessed May 29 2014).

Fru. Developed and published by Through Games. 2016. Xbox One.

"FT [FEMALE TYPE]." Robo Garage (2009).
http://robo-garage.com (accessed June 28 2017).

Fussell, Susan R., Sara Kiesler, Leslie D. Setlock, and Victoria Yew. "How People Anthropomorphize Robots." Paper presented at *3rd ACM/IEEE International Conference on Human Robot Interaction, HRI 2008*. Amsterdam, March 2008.

Galloway, Alexander. "A Response to Graham Harman's 'Marginalia on Radical Thinking'," *An und für sich*: June 3 2012.
https://itself.blog (accessed January 3 2018).

Garber, Megan. "Would You Want Therapy From a Computerized Psychologist?" *The Atlantic* (May 23 2014).

Garland, Alex. *Ex Machina*. United Kingdom and the United States: Universal Studios. 2014. Film.

Garrison, John. *Glass*. Object Lessons. New York: Bloomsbury, 2015.

Germano, William. *Eye Chart*. Object Lessons. New York: Bloomsbury, 2017.

Gibbs, Martin, Marcus Carter, Michael Arnold. "Avatars, Characters, Players and Users: Multiple Identities at/in Play." Paper presented at *OZCHI'12*, Melbourne, Australia, November 26-30 2012.

Gibson, James J. *The Ecological Approach to Visual Perception*. New York: Psychology Press, Taylor and Francis Group, 2015. Originally published 1979.

Gillespie, Craig. *Lars and the Real Girl*. United States and Canada: Metro-Goldwyn-Meyer. 2007. Film.

Goeminne, Gert, and Erik Paredis. "Opening Up the In-Between: Ihde's Postphenomenology and Beyond." *Foundations of Science* 16, no. 2-3 (2011): 101-07.

Gratton, Peter. *Speculative Realism: Problems and Prospects*. London: Bloomsbury, 2014.

Green, Holly. "The Many Deaths of the Kinect." *Paste* (January 10 2018). https://www.pastemagazine.com/ (accessed February 9 2018).

Greenwald, Glenn, Ewen MacAskill, Laura Poitras, Spencer Ackerman and Dominic Rushe. "Microsoft Handed the NSA Access to Encrypted Messages." *The Guardian* (July 12 2013). http://www.theguardian.com/ (accessed January 20 2018).

Grey Walter's Tortoises. 2008. Youtube. https://www.youtube.com/watch?v=1LULRlmXkKo (accessed June 27 2017).

Griffin, David Ray. *Religion and Scientific Naturalism: Overcoming the Conflicts*. Albany NY: State University of New York Press, 2000.

———. *Unsnarling the World-Knot: Consciousness, Freedom, and the Mind-Body Problem*. Berkeley CA: University of California Press, 1998.

Gross, Kenneth. *The Dream of the Moving Statue*. University Park, Pennsylvania: The Pennsylvania State University Press, 2006.

Guizzo, Erico. "Hiroshi Ishiguro: the Man Who Made a Copy of Himself." *IEEE Spectrum* (2010).

Gustafsson, Laura, and Terike Haapoja. *History According to Cattle*. Edited by Laura Gustafsson and Terike Haapoja. Brooklyn, NY: punctum books, 2015.

Halo 4. Developed by 343 Industries. Published by Microsoft Studios. 2012. Xbox 360.

Halsall, Francis. "Art and Guerrilla Metaphysics: Graham Harman and Aesthetics as Ffirst Philosophy." *Speculations: A Journal of Speculative Realism*, no. V (2014): 382-410.

Hampton, Gregory Jerome. *Imagining Slaves and Robots in Literature, Film, and Popular Culture*. Lanham: Lexingon Books, 2015.

Hanson Robotics. "Bina48" (2017). http://www.hansonrobotics.com/ (accessed May 20 2018)

Haraway, Donna. "Anthropocene, Capitalocene, Plantationocene, Chthulucene: Making Kin." *Environmental Humanities* 6 (2015): 159-165.

———. *The Companion Species Manifesto: Dogs, People, and Significant Otherness*. Chicago: Prickly Paradigm Press, 2003.

———. "Situated Knowledges: the Science Question in Feminism and the Privilege of Partial Perspective." *Feminist Studies* 14, no. 3 (Autumn, 1988): 575-599.

Harman, Graham. *Circus Philosophicus*. Winchester, UK: Zero Books, 2010.

———. "The Current State of Speculative Realism." *Speculations: A Journal of Speculative Realism*, no. IV (2013): 22-27.

———. *Guerrilla Metaphysics: Phenomenology and the Carpentry of Things*. Chicago: Open Court, 2005.

———. *Heidegger Explained: From Phenomenon to Thing*. Chicago: Open Court, 2007.

———. *Immaterialism*. Edited by Laurent de Sutter. Theory Redux. Cambridge, UK: Polity Press, 2016.

———. "Materialism Is Not the Solution: on Matter, Form, and Mimesis." *Nordic Journal of Aesthetics* 47 (2015): 94-110.

———. "On Vicarious Causation." In *Collapse* vol. II. Edited by Robin Mackay. Urbanomic, 2007.

———. *Object-Oriented Ontology: A New Theory of Everything*. UK: Pelican Books, 2018.

———. *Prince of Networks: Bruno Latour and Metaphysics*. Melbourne: re.press, 2009.

———. *The Quadruple Object*. Alresford, UK: Zero Books, 2011.

———. "The Road to Objects." *Continent* 1, no. 3 (2011): 171-79.

———. *The Third Table*. Documenta. Kassel, Germany: Hatje Cantz Verlag, 2012.

———. *Tool-Being: Heidegger and the Metaphysics of Objects*. Chicago: Open Court, 2002.

———. "Undermining, Overmining, and Duomining: a Critique." In *ADD Metaphysics*. Edited by Jenna Sutela, 40-51. Aalto, Finland: Aalto University Digital Design Laboratory, 2013.

Harmony AI. Developed by Realbotix. 2017. Android.

Harper, Richard, and Helena M. Mentis. "The Mocking Gaze: the Social Organization of Kinect Use." *Proceedings of the 2013 Conference on Computer Supported Cooperative Work* (2013): 167-80.

Hasse, Cathrine. "Artefacts That Talk: Mediating Technologies as Multistable Signs and Tools." *Subjectivity* 6, no. 1 (2013): 79-100.

Hayler, Matt. *Challenging the Phenomena of Technology: Embodiment, Expertise, and Evolved Knowledge*. Hampshire UK: Palgrave Macmillan. 2015.

Hayles, N. Katherine. "Speculative Aesthetics and Object-Oriented Inquiry (Ooi)." *Speculations: A Journal of Speculative Realism*, no. V (2014): 158-79.

Helgeson, Matt. "Rare Lays Off Staff In Wake of Kinect Sports Rivals." *Gameinformer* (May 19 2014). http://www.gameinformer.com/ (accessed June 27 2017).

Hillis, Ken. "The Sensation of Ritual Space." *Digital Sensations: Space, Identity and Embodiment in Virtual Reality*. Minneapolis: University of Minnesota Press, 1999: 60-89.

Himma, Kenneth Eimar. "Artificial Agency, Consciousness and the Criteria for Moral Agency: What Properties Must an Artificial Agent Have to be a Moral Agent?" *Ethics and Information Technology* 11 (2009): 19-29.

Hoffmeyer, Jesper. "Seeing Virtuality in Nature." *Semiotica* 134, no. 1-4 (2001): 381-398.

Hofstadter, Douglas R. *Gödel, Escher, Bach: An Eternal Golden Braid*. Harmondsworth, England: Penguin Books, 1980.

Holland, Owen. "The First Biologically Inspired Robots." *Robotica* 21 (2003): 351-63.

———. "The Grey Walter Online Archive." University of the West of England (2013). http://www.ias.uwe.ac.uk (accessed September 25 2013).

Howard, Dorothy. "Loving Machines: a De-Anthropocentric View of Intimacy." *Arachne*, no. 00 (2015).

Hsiao, Shih-Wen, Chu-Hsuan Lee, Meng-Hua Yang and Rong-Qi Chen. "User Interface Based on Natural Interaction Design for Seniors." *Computers in Human Behavior* 75 (2017): 147-159.

Hutchens, Jason L. "How to Pass the Turing Test by Cheating." *ACM DL Digital Library*. University of Western Australia, 1996.

Ihde, Don. *Bodies in Technology*. Minneapolis: University of Minnesota Press, 2001.

———. "Introduction: Postphenomenological Research." *Human Studies* 31, no. 1 (2008): 1-9.

———. *Listening and Voice: Phenomenologies of Sound*. Albany, NY: State University of New York Press, 2007.

———. *Postphenomenology and Technoscience: The Peking University Lectures*. Albany, NY: State University of New York Press, 2009.

———. *Postphenomenology: Essays in the Postmodern Context*. Evanston, Illinois: Northwestern University Press, 1993.

———. *Technology and the Lifeworld: From Garden to Earth*. Bloomington: Indiana University Press, 1990.

"[Independent] Chinese Researchers Create Jia Jia - a Super-Lifelike 'Robot Goddess'." University of Science and Technology of China (April 19 2016). http://en.ustc.edu.cn (accessed June 27 2017).

Inglis-Arkell, Esther. "The Very First Robot "Brains" Were Made of Old Alarm Clocks." *io9* (2012). http://io9.com/ (accessed September 25 2013).

"Innovation." Xbox (2014). http://www.xbox.com/en-US/xbox-one/innovation (accessed January 13 2014).

Jones, Gwyneth. *Divine Endurance*. London: George Allen & Unwin, 1984.

Jonze, Spike. *Her*. United States: Warner Bros. Pictures, 2013. Film.

Jordan Wolfson: (Female Figure) 2014. 2014. Youtube. https://www.youtube.com/watch?v=mVTDypgmFCM (accessed June 29 2017).

Jung, Eun Hwa, T. Franklin Waddell and S Shyam Sundar. "Feminizing Robots: User Responses to Gender Cues on Robot Body and Screen." *Conference on Human Factors in Computing Systems Proceedings* (2016): 3107-13.

Juul, Jesper. "The Game, the Player, the World: Looking for a Heart of Gameness." *Digital Games Research Conference: Level Up*. Edited by M. Copier and J. Raessens. Utrecht: Universiteit Utrecht, 2003: 30-47.

Kang, Minsoo. "Building the Sex Machine: The Subversive Potential of the Female Robot." *Intertexts* 9, no. 1 (2005): 5-22.

Kamide, Hiroko and Tatsuo Arai. "Perceived Comfortableness of Anthropomorphized Robots in U.S. and Japan." *International Journal of Social Robotics* 9 (September 2017): 537-543.

karenarchey, "Slavoj Žižek – Objects, Objects Everywhere: a Critique of Object Oriented Ontology." *e-flux conversations* (February 2016). https://conversations.e-flux.com/ (accessed January 3 2018).

Kennedy, Jenny. "Addressing the Wife Drought: Automated Assistants in the Smart Home." Paper presented at *At Home with Digital Media*, Queensland University of Technology, November 2-3 2018.

Khonsari, Michael M. and Mehdi Amiri. *Introduction to Thermodynamics of Mechanical Fatigue*. Boca Raton: CRC Press, 2013.

"kinect." Reddit (2013-2014). http://www.reddit.com/r/Kinect (accessed January 13 2014).

"Kinect on Xbox One." Xbox (2013-2014). http://forums.xbox.com/xbox_forums/xbox_support/xbox_one_support/f/4275.aspx (accessed January 13 2014).

Kinect Sports Rivals. Developed by Rare. Published by Microsoft Studios. 2014. Xbox One.

"Kinect Sports Rivals." Xbox (2014). http://www.xbox.com/he-IL/xbox-one/games/kinect-sports-rivals (accessed May 27 2014).

Kinect Sports Rivals Preseason. Developed by Rare. Published by Microsoft Studios. 2013. Xbox One.

Kinectimals. Developed by Frontier Developments. Published by Microsoft Game Studios. 2010. Xbox 360.

Klepek, Patrick. "Kinect Died in the Uncanny Valley." *Giant Bomb* (May 15 2014). https://www.giantbomb.com/ (accessed February 9 2018).

Koribalski, B., K. Jones, M. Elmouttie and R. Haynes. "Neutral Hydrogen Gas in the Circinus Galaxy." Australia Telescope Outreach and Education, CSIRO. http://outreach.atnf.csiro.au (accessed June 27 2017).

Krahulik, Mike, and Jerry Holkins. "Negotiations." *Penny Arcade* (June 21 2013). http://www.penny-arcade.com (accessed June 26 2017).

Kramer, Jeff. *Hacking the Kinect*. New York: Apress, 2012.

"KSR: Announcement Trailer," Xbox (2014). http://www.xbox.com/en-US/xbox-one/games/kinect-sports-rivals (accessed May 29 2014).

Kubrick, Stanley. *2001: A Space Odyssey*. United States: MGM. 1968. Film.

Kull, Kaveli. "Jakob Von Uexküll: an Introduction." *Semiotica* 134, no. 1-4 (2001): 1-59.

Kurokawa, Megumi, Libby Schweber and Will Hughes. "Client Engagement and Building Design: the View From Actor-Network Theory." *Building Research and Information* 45, no. 8 (2017): 910-925.

Kurzweil, Ray. *How to Create a Mind*. New York: Penguin Books, 2012.

———. *The Singularity is Near*. New York: Viking, 2005.

Kuznets, Lois. *When Toys Come Alive: Narratives of Animation, Metamorphosis and Development*. New Haven: Yale University Press, 1994.

Lang, Fritz. *Metropolis*. Germany: UFA. 1927. Film.

Lash, Scott and Celia Lury. *Global Culture Industry*. Cambridge: Polity Press, 2007.

Latour, Bruno. *Aramis, or the Love of Technology*. Translated by Catherine Porter. Cambridge, MA: Harvard University Press, 1996. Originally published 1993.

———. *Pandora's Hope: Essays on the Reality of Science Studies*. Cambridge, MA: Harvard University Press, 1999.

———. *The Pasteurization of France*. Translated by Alan Sheridan and John Law. Cambridge, MA: Harvard University Press, 1988.

———. *Reassembling the Social: An Introduction to Actor-Network Theory*. Oxford: Oxford University Press, 2005.

———. *Science in Action*. Cambridge, MA: Harvard University Press, 1987.

———. "Technology is Society Made Durable." *Sociological Review Monograph* 38, no. 2 (1991): 103-131.

———. *We Have Never Been Modern*. Translated by Catherine Porter. Hertfordshire: Harvester Wheatsheaf, 1993.

Latour, Bruno, as Jim Johnson. "Mixing Humans and Nonhumans Together: the Sociology of a Door-Closer." *Social Problems* 35, no. 3 (1988): 298-310.

Latour, Bruno, Graham Harman and Peter Erdélyi. *The Prince and the Wolf: Latour and Harman at the LSE*. Winchester, UK: Zero Books, 2011.

Latour, Bruno, and Steve Woolgar. *Laboratory Life: The Social Construction of Scientific Facts*. Beverly Hills: Sage Publications, 1979.

Law, John. "Notes on the Theory of the Actor-Network: Ordering, Strategy and Heterogeneity." *Systems Practice* 5 (1992): 379-93.

Leach, Tessa. "Anthropomorphic Machines: Sensation and Experience in Nonhumans Created to be Like Us". Unpublished thesis, 2018.

———. "Improving on Eyes: Human-Machine Interaction in Cyborg Vision." Unpublished thesis, 2012.

———. "Who Is Their Person? Sex Robot and Change." *Queer-Feminist Science and Technology Studies Forum* 3 (December 2018): 25-39.

Leach, Tessa and Michael Arnold. "How the Kinect Domesticates its Users." Paper presented at *At Home with Digital Media symposium*, Queensland University of Technology, November 3 2017.

"Leap Motion." Leap Motion. 2017. https://www.leapmotion.com (accessed February 16 2018).

Leigh, Stacy. "average americans (that happen to be sex dolls)." stacy leigh. www.stacythearist.com (accessed May 10 2019).

Levy, David. *Love & Sex with Robots: The Evolution of Human-Robot Relationships*. London: Duckworth Overlook, 2009. Originally published 2007.

———. "Roxxxy the "Sex Robot" - Real or Fake." *Lovotics* 1 (2013): 1-4.

The LifeNaut Project. *Bina 48 Meets Bina Rothblatt* (November 27 2014). Youtube. https://www.youtube.com/watch?v=KYshJRYCArE (accessed May 9 2018)

Long, Julia. *Anti-Porn: The Resurgence of Anti-Pornography Feminism*. London: Zed Books, 2012.

Loomis, Roger Sherman. *Celtic Myth and Arthurian Romance*. Chicago: Academy Chicago Publishers, 2005.

Lotman, Juri. "On the Semiosphere." *Sign Systems Studies* 33, no. 1 (2005): 205-29.

Lovelock, James. *Gaia: A New Look at Life on Earth*. Oxford: Oxford University Press, 2000. Originally published 1979.

———. *The Revenge of Gaia*. London: Penguin Books, 2007. Originally published 2006.

Lucas, Gale M., Jonathan Gratch, Aisha King, and Louis-Philippe Morency. "It's Only a Computer: Virtual Humans Increase Willingness to Disclose." *Computers in Human Behavior* 37 (August 2014): 94-100.

Lunning, Frenchy. "Allure and Abjection: the Possible Potential of Severed Qualities." In *Object-Oriented Feminism*. Edited by Katherine Behar, 83-105. Minneapolis: University of Minnesota Press, 2016.

MacDorman, Karl, Sandosh Vasudevan, and Chin-Chang Ho. "Does Japan Really Have Robot Mania? Comparing Attitudes by Implicit and Explicit Measures." *AI & Society* 23, no. 4 (2009): 485-510.

Maddox, Teena. "Sci-fi is turned into reality with technology guru from Minority Report and Iron Man." *TechRepublic* (August 4 2016). https://www.techrepublic.com (accessed December 4 2017).

Maines, Rachel P. *The Technology of Orgasm*. Baltimore: The Johns Hopkins University Press. 1999.

Malin, Brenton J. "Communicating with Objects: Ontology, Object-Orientations, and the Politics of Communication." *Communication Theory* 26 (2016): 236-254.

Manovich, Lev. "Friendly Alien: Object and Interface." *Artifact: Journal of Visual Design* 1, no. 1 (2007): 29-32.

Marino, Mark C. "The Racial Formation of Chatbots." *CLCWeb: Comparative Literature and Culture* 16, no. 5, article 13 (2014).

Markowitz, Judith A., ed. *Robots That Talk and Listen: Technology and Social Impact*. Berlin: De Gruyter, 2015.

Mason, Andrew S. *Plato*. Abingdon, Oxon: Routledge, 2014. Originally published 2010.

Mason, Fran. "Loving the Technological Undead: Cyborg Sex and Necrophilia in Richard Calder's Dead Trilogy." In *The Body's Perilous Pleasures*.

Edited by Michelle Aaron, 108-125. Edinburgh: Edinburgh University Press, 1999.

Mataric, Maja. *The Robotics Primer.* Cambridge, MA: The MIT Press, 2007.

Mattick, Don. "Your Feedback Matters - Update on Xbox One." Xbox (June 19 2013). http://news.xbox.com/2013/06/update (accessed January 15 2014).

Mayr, Ernst. *What Makes Biology Unique?* New York: Cambridge University Press, 2007.

Mazlish, Bruce. *The Fourth Discontinuity: The Co-Evolution of Humans and Machines.* New Haven: Yale University Press, 1993.

McDermott, Drew. "What Matters to a Machine?" In *Machine Ethics.* Edited by Michael Anderson and Susan Leigh Anderson, 88-114. Cambridge: Cambridge University Press, 2011.

McFadyen, Siobhan. "Shocking Tiny Sex Robot Which Looks Like Schoolgirl is on Sale for £770 and Comes Delivered in 'Coffin'." *Mirror* (March 16 2016). http://www.mirror.co.uk (accessed January 21 2018).

Meillassoux, Quentin. *After Finitude* [Après la Finitude (2006)]. Translated by Ray Brassier. New York: Continuum, 2008.

Metacritic. CBS Interactive (2017). www.metacritic.com (accessed May 24 2019).

Microsoft Research. *Behind the eyes of Xbox One Kinect.* 2013. Youtube. https://www.youtube.com/watch?v=JaOlUa57BWs (accessed June 26 2017).

———. *Inside the Brains of Xbox One Kinect.* 2013. Youtube. https://www.youtube.com/watch?v=ziXflemQr3A (accessed June 26 2017).

"Microsoft by the Numbers." Microsoft Story Labs (2013). http://www.microsoft.com (accessed January 24 2014).

Mikelson, Emmy. "Space for Things: Art, Objects, Speculation." In *And Another Thing: Nonanthropocentrism and Art.* Edited by Katherine Behar and Emmy Mikelson, 11-20. Earth, Milky Way: punctum books, 2016.

Miller, Ross. "Kinect for Xbox 360 Review." *engadget* (November 4 2010). https://www.engadget.com (accessed December 4 2017).

Minsky, Marvin. *The Society of Mind*. London: Picador, 1988.

Misses Joust. "Xbox On: Voice Commands for Xbox One." Xbox Forums. 2013-14. http://forums.xbox.com (accessed January 13 2014).

Mol, Annemarie. *The Body Multiple: Ontology in Medical Practice*. Durham, North Carolina: Duke University Press, 2002.

Monk-Turner, Elizabeth. "Gender and Market Opportunity in Japan." *National Journal of Sociology* 11 (1997): 97-105.

Montfort, Nick and Bogost, Ian. *Racing the Beam: The Atari Video Computer System*. Cambridge, MA: MIT Press, 2009.

Mori, Masahiro. *The Buddha in the Robot*. Translated by Charles S. Terry. Tokyo: Kosei Publishing Co., 1981.

Morris, William. "Pygmalion and the Image." *Sacred Texts*. Originally published 1868. http://www.sacred-texts.com (accessed June 27 2017).

Morton, Timothy. *Dark Ecology*. New York: Columbia University Press, 2018.

———. "From Modernity to the Anthropocene: Ecology and Art in the Age of Asymmetry." *International Social Science Journal* 63 (2014): 39-51.

———. "Here Comes Everything: the Promise of Object-Oriented Ontology." *Qui Parle: Critical Humanities and Social Sciences* 19, no. 2 (2011): 163-90.

———. *Hyperobjects: Philosophy and Ecology after the End of the World*. Posthumanities. Minneapolis: University of Minnesota Press, 2013.

Nagel, Thomas. *Mortal Questions*. Cambridge, UK: Cambridge University Press, 1979.

———. "What is it Like to be a Bat?". *The Philosophical Review* 83, no. 4 (1974): 435-50.

"NASA Awards Two Robots to University Groups for R&D Upgrades." NASA (2015). https://www.nasa.gov (accessed June 28 2017).

Nesterov, Nikola, Peter Hughes, Nuala Healy, Niall Sheehy and Neil O'Hare. "Application of Natural User Interface Devices for Touch-Free

Control of Radiological Images During Surgery." Paper presented at *2016 IEEE 29th International Symposium on Computer Based Medical Systems (CBMS)*, Dublin, Ireland, June 20-24 2016.

New Era of Cognitive Computing. (2013).Youtube. https://www.youtube.com/watch?v=h22n80aT2FY (accessed June 27 2017).

Niebisch, Arndt. *Media Parasites in the Early Avant-Garde*. New York: Palgrave MacMillan, 2012.

Nirenberg, S., and C. Pandarinath. "Retinal Prosthetic Strategy with the Capacity to Restore Normal Vision." *Proceedings of the National Academy of Sciences of the United States of America* 109, no. 37 (Sep 11 2012): 15012-7.

Nolan, Jonathan and Lisa Joy. *Westworld*. United States: Warner Bros. Television, 2016. Television series.

"The Nonhuman Turn Conference", Center for 21st Century Studies (blog). https://c21uwm.wordpress.com/the-nonhuman-turn/ (accessed January 17 2018)

Norman, Donald. "Affordance, Conventions and Design." *Interactions* 3 (May/June 1999). http://interactions.acm.org (accessed March 18 2018).

———. *The Design of Everyday Things: Revised and Extended Edition*. New York: Basic Books, 2013. Originally published 1988.

———. "Natural User Interfaces are not Natural." *Interactions* 17, no. 3 (2010).

Nöth, Winfried. "Semiosis and the Umwelt of a Robot." *Semiotica* 134, no. 1-4 (2001): 695-99.

"Object lessons." objectsobjectsobjects.com (accessed June 22 2017).

"Objectùm-Sexuality Internationale." http://www.objectum-sexuality.org/ (accessed June 27 2017).

Olivares, Jonathan, interviewing Graham Harman. "The Metaphysics of Logos: Decoding Branded Objects with Philosopher Graham Harman." *032c* (May 17 2018). https://032c.com (accessed May 28 2018).

Operation: Pedopriest. Developed by Molleindustria. 2007. PC.

O'Rourke, Michael. "Girls Welcome!!!" *After the "Speculative Turn": Realism, Philosophy, and Feminism*. Edited by Katerina Kolozova and Eileen A. Joy, 159-197. Earth: punctum books, 2016.

Osborn, Henry Fairfield. "The Evolution of Mammalian Molars to and from the Tritubercular Type." *The American Naturalist* 22, no. 264 (December 1888): 1067-1079.

Ovid. "Metamorphoses." *The Ovid Collection*. Translated by A.S. Kline (2010). http://ovid.lib.virginia.edu/trans/Ovhome.htm (accessed June 27 2017).

Panzarino, Matthew. "The New Xbox One Kinect Tracks Your Heart Rate, Happiness, Hands and Hollers." *The Next Web* (May 22 2013). http://thenextweb.com (accessed June 26 2017).

Parikka, Jussi. *An Alternative Deep Time of the Media: a Geologically Tuned Media Ecology*. 2013. Vimeo.
http://jussiparikka.net (accessed August 22 2013).

———. *Insect Media*. Posthumanities. Minneapolis: University of Minnesota Press, 2010.

Parkin, Simon. "Kinect Sports Rivals Review: Warm Down." *Eurogamer* (April 8 2014). http://www.eurogamer.net (accessed May 29 2014).

"Parliament of Things." Parliament of Things (2017). https://theparliamentofthings.org (accessed December 21 2017).

Pechenik, Jan A. *Biology of the Invertebrates*. Fifth edition. Boston: McGraw Hill Higher Education, 2005.

Perkowitz, Sidney. *Digital People: From Bionic Humans to Androids*. Washington DC: Joseph Henry Press, 2004.

Petersen, Steve. "Is it Good for Them Too? Ethical Concern for the Sexbots." *Robot Sex: Social and Ethical Implications*. Edited by John Danaher and Neil McArthur, 155-172. Cambridge, MA: The MIT Press, 2017.

Petit, Carolyn. "Kinect Sports Rivals Review: Championing Mediocrity." *Gamespot* (April 7 2014).
http://www.gamespot.com (accessed May 29 2014).

Phan, Thao. "The Materiality of the Digital and the Gendered Voice of Siri." *Transformations* 29 (2017): 23-33.

Pickering, Andrew. *The Cybernetic Brain: Sketches of Another Future*. Chicago: University of Chicago Press, 2010.

Plunkett, Luke. "That Xbox One Reveal Sure Was a Disaster, Huh?" *Kotaku* (May 21 2013). http://kotaku.com (accessed May 28 2013).

"Realbotix." Realbotix (2016). https://realbotix.com/ (accessed May 25 2016).

"RealDoll." RealDoll (2015). https://secure.realdoll.com/ (accessed February 18 2016).

Revesz, Rachael. "Paedophiles 'Could be Prescribed Child Sex Dolls' to Prevent Real Attacks, Says Therapist." *Independent* (August 2 2017). http://www.independent.co.uk (accessed January 8 2018).

Richardson, Kathleen. "Sex Robot Matters." *IEEE Technology and Society Magazine* (June 2016).

———. "Urgent – Why We Must Campaign to Ban Sex Dolls & Sex Robots." Campaign Against Sex Robots (October 3 2017). https://campaignagainstsexrobots.org (accessed May 26 2018).

Rise of Nightmares. Developed by Sega. Published by Sega. 2011. Xbox 360.

Riskin, Jessica. "Machines in the Garden." *Republics of Letters: A Journal for the Study of Knowledge, Politics and the Arts* 1, no. 2 (April 30 2010).

Rivington, James. "Microsoft Kinect for Xbox 360 review." *Techradar* (November 4 2010). http://www.techradar.com/reviews (accessed December 4 2017).

Robertson, Jennifer. "Gendering Humanoid Robots: Robo-Sexism in Japan." *Body & Society* 16, no. 2 (2010): 1-36.

———. "Gendering Robots: Posthuman Traditionalism in Japan." In *Recreating Japanese Men*. Edited by Sabine Frühstück and Anne Walthall, 311-332. Berkeley: University of California Press, 2011.

Rosenberger, Gregg. *A Place for Consciousness: Proving the Deep Structure of the Natural World*. Oxford: Oxford University Press, 2004.

Rosenberger, Robert. "Multistability and the Agency of Mundane Artifacts: From Speed Bumps to Subway Benches." *Human Studies* 37 (2014): 369-92.

———. "The Sudden Experience of the Computer." *AI & Society* 24 (2009): 173-180.

Rosenberger, Robert, and Peter-Paul Verbeek, eds. *Postphenomenological Investigations: Essay on Human-Technology Relations*. Lanham, Maryland: Lexington Books, 2015.

Rothstein, Adam. *Drone*. Object Lessons. New York: Bloomsbury, 2015.

Roy, Deb. "Semiotic Schemas: a Framework for Grounding Language in Action and Perception." *Artificial Intelligence* 167 (2005): 170-205.

Russell, Kyle. "People are Worried Microsoft's new Xbox Will be Able to Spy on You." *Business Insider Australia* (May 29 2013). https://www.businessinsider.com.au (accessed February 22 2018).

Ryse: Son of Rome. Developed by Crytek. Published by Microsoft Studios. 2013. Xbox One.

Salih, Sara. "On Judith Butler and Performativity." In *Sexualities and Communication in Everyday Life: A Reader*. Edited by Karen E. Lovaas and Mercilee M. Jenkins, 55-67. Thousand Oaks, CA: SAGE Publications, 2007.

Salter, Ben. "E3: Xbox One's Kinect Knows Too Much." Xbox.MMGN.com (June 14 2013). http://mmgn.com (accessed June 19 2013).

Schodt, Frederik L. *Inside the Robot Kingdom: Japan, Mechatronics and the Coming Robotopia*. Tokyo: Kodansha International Ltd., 1988.

Schouten, Peer. "The Materiality of State Failure: Social Contract Theory, Infrastructure and Governmental Power in Congo." *Millennium: Journal of International Studies* 41, no. 3 (2013): 553-574.

Scobie, A., and J. Taylor. "Perversions Ancient and Modern: I. Agalmatophilia, the Statue Syndrome." *Journal of the History of the Behavioral Sciences* 11, no. 1 (1975): 49-54.

Scott, Ridley. *Blade Runner*. United States: Warner Bros. 1982. Film.

Searle, John R., "Minds, Brains, and Program." In *The Turing Test: Verbal Behaviour as the Hallmark of Intelligence*. Edited by Stuart M. Shieber, 201-224. Cambridge, MA: MIT Press, 2004.

Searle, John, and David Chalmers. "'Consciousness and the Philosophers': an Exchange." *The New York Review of Books*. (May 15 1997).

Sebeok, Thomas A. "Biosemiotics: its Roots, Proliferation, and Prospects." *Semiotica* 134, no. 1-4 (2001): 61-78.

Sebeok, Thomas A., and J. Umiker-Sebeok. *The Semiotic Web 1991*. Berlin: Mouton de Gruyter, 1992.

Selinger, Evan. "Normative Judgement and Technoscience: Nudging Ihde, Again." *Techné* 12, no. 2 (2008): 120-25.

Shah, Agam. "New Kinect for Windows to Improve Human Interaction with Computers." *Computerworld* (May 23 2013). https://www.computerworld.com (accessed February 22 2018).

Sharkey, Noel, Aimee van Wynsberghe, Scott Robbins and Eleanor Hancock. *Our Sexual Future with Robots*. The Hague: Foundation for Responsible Robotics, 2017. PDF.

Shaviro, Steven. *The Universe of Things*. Minneapolis: University of Minnesota Press, 2014.

Shaw, Ian G. R., and Katherine Meehan. "Force-Full: Power, Politics and Object-Oriented Philosophy." *Area* 45, no. 2 (2013): 216-22.

Sherwood, Harriet. "Robot Priest Unveiled in Germany to Mark 500 Years Since Reformation." *The Guardian* (May 30 2017). https://www.theguardian.com (accessed December 4 2017).

SHRDLU. Developed by Terry Winograd and Greg Sharp. Published by Sun Microsystems, Inc. 1999. Windows 7.

"SHRDLU." http://hci.stanford.edu/winograd/shrdlu/ (accessed January 22 2018).

Siciliano, Leon and Reuters. "A Japanese Company just Unveiled a Robot Priest that will read Scriptures at Buddhist Funerals." *Business Insider* (August 24 2017). http://www.businessinsider.com (accessed June 18 2018).

Silverstone, Roger, and Leslie Haddon. "Design and the Domestication of ICTs: Technical Change and Everyday Life." In *Communication by Design: The Politics of Information and Communication Technologies*. Edited by R. Silverstone and R. Mansell, 44-74. Oxford: Oxford University Press, 1996.

Simonds, Will. "Xbox One Will Know Your Face, Voice, and Heartbeat." *Abine* (May 21 2013).
https://www.abine.com/blog (accessed February 22 2018).

Simons, Geoff. *Robots*. London: Cassell, 1992.

Sirani, Jordan. "Microsoft Won't Offer Free Kinect Adaptor With Xbox One X Upgrade." *IGN* (October 10 2017).
http://au.ign.com (accessed December 7 2017).

Skrbina, David. *Mind that Abides: Panpsychism in the New Millennium*. Edited by David Skrbina. Amsterdam: John Benjamins Publishing Company, 2009.

———. *Panpsychism in the West*. Cambridge, MA: MIT Press, 2005.

Slabaugh, Brett. "Xbox One's Kinect Can Actually Be Turned Off." *The Escapist* (May 30 2013). http://www.escapistmagazine.com (accessed January 20 2018).

Søraa, Roger Andre. "Mechanical Genders: How do Humans Gender Robots?" *Gender, Technology and Development* 21, no. 1-2 (2017): 99-115.

Sparrow, Robert. "Robots, Rape, and Representation." *International Journal of Social Robotics* 9 (September 2017):465-477.

Sparrow, Tom. *The End of Phenomenology*. Edinburgh: Edinburgh University Press, 2014.

Spencer, Phil. "Delivering More Choices for Fans." Xbox Wire (blog) (May 13 2014).
http://news.xbox.com (accessed June 26 2017).

Spielberg, Steven. *Minority Report*. United States: 20th Century Fox and Dreamworks Pictures, 2002.

Stock, Wolfgang G., and Mechtild Stock. *Handbook of Information Science*. Translated by Paul Becker. Berlin: De Gruyter, 2013.

Strikwerda, Litska. "Legal and Moral Implications of Child Sex Robots." In *Robot Sex: Social and Ethical Implications*. Edited by John Danaher and Neil McArthur, 133-152. Cambridge, MA: MIT Press, 2017.

Suzuki, Miwa. "In Japan, Aibo Robots get their own Funeral." *The Japan Times* (May 1 2018). https://www.japantimes.co.jp (accessed June 18 2018).

Tashev, Ivan. "Recent Advances in Human-Machine Interfaces for Gaming and Entertainment." *International Journal of International Technology and Security* 3, no. 3 (2011): 69-76.

Tatsumi, Takayuki. *Full Metal Apache: Transactions between Cyberpunk Japan and Avant-Pop America*. Durham: Duke University Press, 2006.

Tavinor, Grant. "Definition of Videogames." *Contemporary Aesthetics* 6 (2008): 1-17.

Toffoletti, Kim. "Barbie: a Posthuman Prototype." *Cyborgs and Barbie Dolls*. London: I.B. Tauris, 2007.

Trachtenberg, Joshua. *Jewish Magic and Superstition: A Study in Folk Religion*. Philadelphia, Pennsylvania: University of Pennsylvania Press, 2004. Google Ebook. Originally published 1939.

"Truecompanion.com." True Companion (2016). http://www.truecompanion.com/home.html (accessed May 25 2016).

TruecompanionLLC. *Roxxxy Sex Robot Meets Dr Oz* (2018). Youtube. https://www.youtube.com/watch?v=540jv33Nm9Y (accessed June 18 2018).

Trung, Le. "When Science Meets Beauty." Project Aiko (2007-2013). http://www.projectaiko.com/index.html (accessed June 29 2017)

Turkle, Sherry. *Alone Together*. New York: Basic Books, 2011.

———. Life on the Screen. New York: Simon & Schuster Paperbacks, 1995.

Uexküll, Jakob von. *A Foray into the Worlds of Animals and Humans: with a Theory of Meaning*. Translated by Joseph D. O'Neill. Posthumanities. Minneapolis: University of Minnesota Press (2010).

———. "A Stroll Through the World of Animals and Men: a Picture Book of Invisible Worlds." *Semiotica* 89, no. 4 (1992): 319-91.

———. "An Introduction to Umwelt." *Semiotica* 134, no. 1-4 (2001): 107-10.

———. *Theoretical Biology*. Translated by D. L. MacKinnon. Edinburgh: The Edinburgh Press, 1926.

———. *Umwelt and Innenwelt of the Tick*. Berlin: Verlag von Julius Springer, 1909.

Uexküll, Thure von. "Introduction: the Sign Theory of Jakob von Uexküll." *Semiotica* 89, no. 4 (1992): 279-315.

Ulanoff, Lance. "Xbox 360: A Tale From the Red Ring of Death." *PC Magazine* (July 28 2010.) http://www.pcmag.com (accessed June 27 2017).

Van Atten, Mark. *Essays on Gödel's Reception of Leibniz, Husserl, and Brouwer*. Switzerland: Springer International Publishing, 2015.

Van den Eede, Yoni. "The Mediumness of World: A Love Triangle of Postphenomenology, Media Ecology, and Object-Oriented Philosophy." In *Postphenomenology and Media: Essays on Human-Media-World Relations*. Edited by Yoni van den Eede, Stacey O'Neal Irwin, and Galit Wellner, 229-250. Lanham: Lexington Books, 2017.

Verbeek, Peter-Paul. "Expanding Mediation Theory." *Foundations of Science* 17 (2012): 391-395.

———. "The Morality of Things: a Postphenomenological Inquiry." In *Postphenomenology: A Critical Companion to Ihde*. Edited by Evan Selinger, 117-128. Albany, New York: State University of New York Press, 2006.

———. *What Things Do: Philosophical Reflections on Technology, Agency and Design* [De daadkracht der Dingen: Over techniek, filosofie en vormgeving]. Translated by Robert P. Crease. University Park Pennsylvania: Pennsylvania State University Press, 2005. Originally published 2000.

Vidal, Denis. "Anthropomorphism or Sub-Anthropomorphism? An Anthropological Approach to Gods and Robots." *Journal of the Royal Anthropological Institute* 13, no. 4 (2007): 917-33.

Vonnegut, Kurt. *Slaughterhouse Five*. New York: Bantam Doubleday Dell Publishing Group, 1998. Originally published 1969.

"Voyager 2: In Depth." Solar System Exploration, NASA Science. https://solarsystem.nasa.gov (accessed June 22 2017).

"Voyager Goes Interstellar." Voyager The Interstellar Mission. https://voyager.jpl.nasa.gov/ (accessed June 22 2017).

Wallach, Wendell, and Colin Allen. *Moral Machines: Teaching Robots Right from Wrong*. Oxford: Oxford University Press, 2009.

Walter, William Grey. "A Machine That Learns." *Scientific American* 185 no. 2 (August 1951).

———. "An Imitation of Life." *Scientific American* 182, no. 5 (1950): 42-45.

———. *The Living Brain*. London: Gerald Duckworth & Co. Ltd., 1953.

Waytz, Adam, Nicholas Epley, and John T. Cacioppo. "Social Cognition Unbound: Insights into Anthropomorphism and Dehumanization." *Current Directions in Psychological Science* 19, no. 1 (2010): 58-62.

Webber, Sarah, Marcus Carter, Wally Smith and Frank Vetere. "Interactive Technology and Human-Animal Encounters at the Zoo." *International Journal of Human-Computer Studies* 98 (2017): 150-168.

Webster, Andrew. "Kinect Used to Control 83-Year-Old, Four-Story High Organ in Australia." *The Verge* (April 9 2012). http://www.theverge.com (accessed January 20 2018).

Wii Fit. Developed by Nintendo EAD Group No. 5. Published by Nintendo. 2008. Nintendo Wii.

Wii Sports. Developed by Nintendo EAD. Published by Nintendo. 2006. Nintendo Wii.

Williams, George Christopher. *Adaptation and Natural Selection*. New Jersey and Chichester, UK: Princeton University Press, 1966.

Winner, Langdon. "Do Artifacts Have Politics?" *The Whale and the Reactor: a Search for Limits in an Age of High Technology*. Chicago: University of Chicago Press, 1986.

———. "Upon Opening the Black Box and Finding it Empty: Social Constructivism and the Philosophy of Technology," *Science, Technology and Human Values* 18, no. 3 (Summer 1993): 362-378.

Winograd, Terry. *Five Lectures on Artificial Intelligence*. Springfield, VA: National Technical Information Service, 1974.

———. "Procedures as a Representation for Data in a Computer Program for Understanding Natural Language." Doctoral dissertation, Massachusetts Institute of Technology, 1971.

———. "Understanding Natural Language." *Cognitive Psychology* 3 (1972): 1-191.

———. "What Does it Mean to Understand Language?" *Cognitive Science* 4 (1980): 209-41.

Winograd, Terry and Fernando Flores. *Understanding Computers and Cognition*. Reading, MA: Addison Wesley Publishing Company, Inc., 1987.

Wood, Gaby. *Edison's Eve: A Magical History of the Quest for Mechanical Life*. New York: Alfred A. Knopf, 2002.

———. *Living Dolls: A Magical History of the Quest for Mechanical Life*. London: Faber and Faber Limited, 2002.

Woolgar, Steve. "Configuring the User." *Sociological Review* 38 (1990): 58-99.

Wolfendale, Peter. "The Noumenon's New Clothes (Part 1)." *Speculations III* (2012): 290-366.

Wosk, Julie. *My Fair Ladies: Female Robots, Androids and Other Artificial Eves*. New Jersey: Rutgers University Press, 2015.

Xbox On. 2013. Youtube. http://www.youtube.com/watch?feature=player_embedded&v=Fm95LxARCFE (accessed January 13 2014).

"XBOX ONE." Reddit (2013-2014). http://www.reddit.com/r/XboxOne (accessed January 13 2014).

Xbox One Voice Commands Don't Work?!?! Xbox One FAIL (2013). Youtube. http://www.youtube.com/watch?feature=player_embedded&v=JSv7iEpAE58 (accessed January 13 2014).

"Xbox One: What it is." Xbox. 2013. http://www.xbox.com (accessed June 2 2013).

Yalom, Irvin D. *Love's Executioner and Other Tales of Psychotherapy*. New York: Basic Books, 1989.

Zahavi, Dan. "The End of What? Phenomenology vs. Speculative Realism." *International Journal of Philosophical Studies* 24, no. 3 (2016): 289-309.

Zhang, Zheng. "The Changing Landscape of Literacy Curriculum a Sino-Canada Transnational Education Programme: an Actor-Network Theory Informed Case Study." *Journal of Curriculum Studies* 48, no. 4 (2016): 547-564.

Zielinski, Siegfried. *Deep Time of the Media*. Cambridge, MA: The MIT Press, 2006.

www.ingramcontent.com/pod-product-compliance
Lightning Source LLC
Chambersburg PA
CBHW071658160426
43195CB00012B/1507